Edward Joseph Lowe

A natural History of British Grasses

Edward Joseph Lowe

A natural History of British Grasses

ISBN/EAN: 9783743331341

Manufactured in Europe, USA, Canada, Australia, Japa

Cover: Foto ©berggeist007 / pixelio.de

Manufactured and distributed by brebook publishing software (www.brebook.com)

Edward Joseph Lowe

A natural History of British Grasses

CONTENTS

AND

LIST OF COLOURED PLATES.

	Plate.	Page.		Plate.	Page.
Agrostis alba	xvii B	59	Bromus erectus	li	157
canina	xvi B	55	arvensis	lvi	167
setacea	xvii A	57	commutatus	lv	165
spica-venti	xviii B	63	diandrus	lvii B	171
vulgaris	xviii A	61	maximus	lviii	173
Aira alpina	xxi A	69	mollis	lvii A	169
cæspitosa	xx	67	sterilis	liii	161
canescens	xxiii A	75	secalinus	liv	163
caryophyllea	xxi B	71	Calamagrostis epigejos	xv A	49
flexuosa	xxii	73	lanceolata	xv B	51
præcox	xxiii B	77	stricta	xvi A	53
Alopecurus agrestis	iv	11	Catabrosa aquatica	xix	65
alpinus	iii B	9	Cynodon dactylon	lxxi	217
bulbosus	v A	13	Cynosurus cristatus	xliv A	139
fulvus	v B	15	echinatus	xliv B	141
geniculatus	vi	17	Dactylis glomerata	xli B	133
pratensis	iii A	7	Digitaria humifusa	lxxiii	221
Ammophila arundinacea	viii A	23	sanguinalis	lxxii	219
Anthoxanthum odoratum	i	3	Festuca elatior	xlvi	145
Arrhenatherum avenaceum			bromoides	l A	153
	xxviii	89	gigantea	xlvii	147
Avena fatua	lix	175	ovina	l B	155
flavescens	lxii A	183	pratensis	xlv	143
pratensis	lx A	177	sylvatica	xlix	151
pubescens	lx B	179	uniglumis	xlviii	149
strigosa	lxi	181	Gastridium lendigerum	xi B	39
Brachypodium pinnatum	lxvi B	201	Hierochloe borealis	xxix A	91
sylvaticum	lxvi A	199	Holcus lanatus	xxvii	87
Briza media	xlii	135	mollis	xxvi	85
minor	xliii	137	Hordeum maritimum	xliv A	191
Bromus asper	lii	159	murinum	lxiii B	189

CONTENTS AND LIST OF COLOURED PLATES (Continued).

	Plate.	Page.
Hordeum pratense	lxiii A	187
sylvaticum	lxii B	185
Knappia agrostidea	lxix A	211
Koeleria cristata	xxix B	93
Lagurus ovatus	xi A	37
Lepturus incurvatus	lxviii B	209
Lolium multiflorum	lxvii B	205
perenne	lxvii A	203
temulentum	lxviii A	207
Melica nutans	xxv A	81
uniflora	xxv B	83
Milium effusum	xii	41
Molinia cœrulea	xxiv	79
Nardus stricta	ii	5
Panicum crus-galli	xxx B	97
Phalaris arundinacea	vii B	21
canariensis	vii A	19
Phleum alpinum	ix A	27
arenarium	x B	35
asperum	ix B	29
boehmeri	x A	33
michelii		31
pratense	viii B	25
Phragmitis communis	lxxiv	223
Poa aquatica	xxxii	103
alpina	xxxix B	125

	Plate.	Page.
Poa annua	xl B	129
bulbosa	xxxix A	123
compressa	xxxvii B	119
distans	xxxiii	105
fluitans	xxxiv A	107
loliacea	xxxvii A	117
maritima	xxxiv B	109
nemoralis	xl A	127
pratensis	xxxvi	115
procumbens	xxxv A	111
rigida	xxxv B	113
trivialis	xxxviii	121
Polypogon littoralis	xiv B	47
monspeliensis	xiv A	45
Sesleria cœrulea	xxx A	95
Setaria glauca		102
verticillata	xxxi A	99
viridis	xxxi B	101
Spartina alterniflora	lxx	215
stricta	lxix B	213
Stipa pennata	xiii	43
Triodia decumbens	xli A	131
Triticum caninum	lxv B	197
junceum	lxiv B	193
repens	lxv A	195

BRITISH GRASSES.

GRAMINEÆ. Jussieu.

FLORETS mostly perfect, yet occasionally imperfect, or even without stamens or pistil. One, two, or more imbricated on a common axis or rachis situated within an involucre, called a *calyx* by Linnæus, consisting of one or two (rarely none) valves or *glumes*, the whole constituting a *spikelet*.

Perianth, (called corolla by Linnæus,) glumaceous, the fertile florets generally consisting of two dissimilar *glumellas* or *valvelets*. The exterior or lower one simple, mostly keeled or having a midrib, the interior or upper one having two lateral or dorsal nerves: occasionally one or even both are wanting.

Stamens hypogynous, usually three, but either one, two, three, four, five, six, seldom indefinite.

Anthers bicelled, attached by their back near the middle, versatile.

Ovary superior, single-celled with one *ovule*, having mostly two (more rarely one, or even none) diminutive hypogynous scales, called *lodicules* or abortive stamens.

Styles usually two, which are simple or bifid, more rarely one or three.

Stigmas mostly plumose.

Pericarp closely incorporated with the seed.

Embrio lenticular, external, situated at the base of the farinaceous albumen.

The culms or stems generally fistulose, mostly simple, herbaceous, and knotted. Occasionally branched, seldom shrubby. Hollow mostly, being closed at the joints.

Leaves, a single one to each node, having a sheath slit longitudinally on one side, and frequently possessing a membranous appendage at the summit, called a *ligule*.

The flowers, which are small, are solitary or in spikelets, which are panicled or spiked.

ANTHOXANTHUM ODORATUM.

Linnæus. Parnell. Vahl. Martyn. Stillingfleet. Koch. Smith. Sinclair. Schreber. Poiteau and Turpin. Hooker. Greville. Arnott. Lindley. Willdenow. Curtis. Leers.

PLATE I.

The Sweet-scented Vernal Grass.

Anthoxanthum—A yellow flower, (from the Greek.)   *Odoratum*—Sweet.

Anthoxanthum. *Linnæus.*—Having two stamens and two styles. A spiked panicle. The spikelet having one central fertile floret. Two glumes. Four glumellas, the two inner ones perfect florets, and awnless; the two exterior ones neuter florets, larger, and awned. Only one British species.

THE pleasant odour that is so very powerful in our hay-fields, whilst the grass is drying, owes much of its scent to the present species, a scent very similar to that of the Woodruff, (*Asperula odorata.*) It springs up early, and is a true permanent pasture grass, and is to be met with almost everywhere both on strong and light soils.

A common species throughout Europe, and in the more northerly parts of North America.

Panicle upright, form ovate oblong, with short hairy branches, length an inch and a half. Spikelets in form ovate-lanceolate, large, erect, four or five together, one awned floret. Calyx two remarkably unequal glumes, which are hairy; the larger glume is three-ribbed, the outer smaller glume destitute of lateral ribs. Floret consisting of two paleæ, oblong in shape, hairy, brownish, the paleæ equal in size, being half the length of the larger glume, and having two awns very dissimilar in length.

Stamens consisting of two, which extend beyond the spikelet. Anthers oblong, and notched at the extremities. Styles brief and smooth. Ovarium oblong. Stigmas long, downy, and extending beyond the summit. Seed solitary, naked, and pointed at each extremity. Stem circular, striated, very smooth, having two or three hairy striated sheaths; the upper sheath extending beyond its leaf. Joints long and distant. Leaves flat, pale green, ribbed, hairy both above and beneath. Inflorescence simple, panicled. Length of the Grass from twelve to eighteen inches. Root perennial, fibrous.

Flowers in the middle of April, and ripens its seeds in the middle of June.

A valuable agricultural Grass.

I am indebted to Mr. Joseph Sidebotham, of Manchester, and to Dr. Wilson, of Nottingham, for specimens of this species.

The illustration is from Dr. Wilson's specimen.

NARDUS STRICTA.

NARDUS STRICTA.

Linnæus. Hooker and Arnott. Willdenow. Vahl. Leers. Koch. Parnham. Knapp. Smith. Cavanilles. Sowerby. Sinclair. Greville. Lindley. Hudson. Schreber. Withering. Martyn. Schrader.

PLATE II.

The Mat Grass.

Nardus—Odoriferous, (from the Greek.)   *Stricta*—Upright.

Nardus. *Linnæus.*—Spikelets simple, on one side of the rachis only. Glumes wanting. Glumellas two; the outer one keeled. Stigma elongated, filiform, and papillose. Stamens three in number. Confined to a solitary British species.

A MOST abundant Grass on moors and sandy wastes. There is a rush-like character in its leaves, which are rigid, harsh, and forming thick tufts which continue all winter.

It is of no use for agricultural purposes, cattle rejecting it if they can find other food.

Common throughout Scotland, England, Ireland, Lapland, Norway, Sweden, and Germany: it is also found in the most northerly portions of North America.

Spike single; spikelets single-flowered, lanceolate in form, deposited in two rows on one side of the rachis only; the opposite side of rachis naked. Without a calyx. Glumes none. Florets consisting of two paleæ, the exterior one tipped with a brief rough awn; the inner palea shorter, entire, membranous, and linear-lanceolate in form. Anthers oblong. Ovarium also oblong and slender. Style only one.

Stigma one, elongated, filiform, papillose. Seed solitary, linear, and pointed at each extremity. Stem erect, mostly smooth, having four or five leaves, with sheaths that are also smooth and striated, the uppermost sheath extending beyond its leaf, whilst the lower sheaths are considerably shorter than their leaves. Joints placed near the base of the stem. Leaves rolled inwards, edges rough, bristle-shaped, striated, harsh and rigid, and suddenly branching off at a right angle. Inflorescence spiked. Spike upright and close. Length of Grass from five to eight inches. Root perennial, having many stout downy fibres.

Comes into flower at the beginning of July, and ripens seed at the beginning of August.

I am indebted to Mr. Joseph Sidebotham, of Manchester, for the specimen illustrated.

ALOPECURUS PRATENSIS.

LINNÆUS. PARNELL. SCHREBER. HOOKER AND ARNOTT. KOCH. LEERS. GREVILLE. CURTIS. SINCLAIR. LINDLEY. SMITH. KNAPP. HUDSON. SCHRADER. WITHERING. WILLDENOW. MARTYN. GRAVES.

PLATE III.—A.

Gramen alopecuroides majus,　　　　MORISON. GERARDE.

The Meadow Fox-tail Grass.

Alopecurus—Fox-tail, (from the Greek.)　　*Pratensis*—Meadow.

ALOPECURUS. *Linnæus.*—Inflorescence a thick, close-set panicle, which is spiked. The spikelets are laterally compressed. Two almost equal glumes, their base mostly connate; membranaceous, and of the same length as the floret. A solitary glumella, having a dorsal awn above the base.

ONE of the most valuable agricultural Grasses; cattle are exceedingly fond of it, and being good in quality, and an early species. It flourishes best in meadows which have been drained, on a rich clayey soil.

Exceedingly common in Great Britain, and is found also in Russia, Sweden, Norway, Denmark, Lapland, Holland, France, Germany, Italy, and America. In the latter country it is however supposed to have been introduced.

Panicle upright, varying from an inch to two inches long, nearly cylindrical in form, having small branches arranged all round the rachis. Spikelets ovate in form, erect, imbricated, numerous. Calyx consisting of two equal-length glumes, acute and jointed at the base, pale green lateral ribs and keels, which are fringed. Floret consisting of a solitary ovate-oblong palea, with two green ribs on either side; having a lengthened slender dorsal awn arising almost immediately above the base. Upper portion of the keel hairy. Anthers conspicuous, yellow in colour. Styles joined together. Stigmas divided, slender, and downy. Seed ovate. Stem erect, circular, polished, and striated, carrying four or five leaves, with smooth inflated sheaths. Joints smooth. Leaves flat, acute, both surfaces usually rough. Inflorescence compound, branched. Panicle upright, from one to two inches long; cylindrical-oblong, compact, with short branches placed all round the rachis. Length varying from one to three feet, according to circumstances. Root perennial and fibrous.

Flowers through April, May, and June, and ripens its seed in July and August.

Dr. Parnell gives the following distinctions in his "Grasses of Scotland," from other species in the same genus:—

"1st.—From *A. geniculatus* in *upper* leaf being only half the length of its sheath; *awn* extending more than half its length beyond the palea; *palea* conical, with four *distinct* broad green ribs; *glumes* more acute, and of a different shape, whereas in *A. geniculatus* the upper leaf is nearly the same length as its shield; *palea* obtuse, with four *indistinct* green ribs, tinged at apex with purple.

2nd.—From *A. fulvus* in the *awn* of *A. fulvus* not extending beyond the palea.

3rd.—From *A. agrestis* in the stems and sheaths being smooth, in *A. agrestis* they are rough.

4th.—From *A. alpinus* in the panicle being longer; in *A. alpinus* it is not an inch long; also differs in the position of the *awn*."

My thanks are due to Dr. Wilson, of Nottingham, and to Mr. Joseph Sidebotham, of Manchester, for the illustrated specimens.

ALOPECURUS ALPINUS.

Smith. Hooker and Arnott. Don. Knapp. Parnell. Lindley.

PLATE III.—B.

Alopecurus ovatus,   Knapp.

The Alpine Fox-tail Grass.

Alopecurus—Fox-tail.   *Alpinus*—Alpine.

A RARE northern Grass, growing in marshy situations on mountains at an elevation of three thousand eight hundred feet above the sea. Sheep will feed upon the lower leaves, rejecting the stem of this Grass. It does not appear to be found below the elevation of two thousand five hundred feet.

From the circumstance that *Alopecurus alpinus* will not grow except on high mountains it is useless as an agricultural Grass.

In Great Britain it is peculiar to the Highlands of Scotland, about Loch-na-Gar, Clova Mountains, Canlochen, Glen Prosen, Ben Lawers, and Loch Leo. Found also in Greenland, in Spitzbergen, and in the north of British America.

Panicle upright, not an inch long, oblong, soft and silky. Spikelets upright, oval, placed all round the rachis, of one awned floret of the same length as the calyx. Calyx consisting of two acute hairy glumes of the same size, three-ribbed. Floret consisting of one palea, with two ribs on either side, and sometimes furnished with a slender dorsal awn. Filaments three and slender. Anthers protruding, and notched at the extremities. Styles united and short. Stigmas two, lengthy and feathery. Seeds ovate. Stem upright, smooth, circular, bent at the base. Stem carrying four leaves, whose sheaths are smooth and striated, the uppermost sheath extending beyond its leaf. Joints smooth. Leaves flat and broad, the inner surface and margin rough.

Inflorescence panicled. Length from nine inches to a foot. Root perennial, creeping, having lengthened fibres.

Flowers in July, and ripens seed at the end of August.

The illustration was forwarded by Mr. Joseph Sidebotham, of Manchester, having been gathered on the Clova Mountains.

ALOPECURUS AGRESTIS.

Linnæus. Willdenow. Smith. Martyn. Hooker and Arnott.
Leers. Knapp. Schrader. Schreber. Ehrhart. Sinclair.

PLATE IV.

Alopecurus myosuroides,   Hudson. Curtis.

The Slender Fox-tail Grass.

Alopecurus—Fox-tail.   *Agrestis*—A field.

A USELESS agricultural Grass, indeed cattle refuse to feed upon it; when once it takes possession of a field it is difficult to eradicate, and causes much trouble to farmers when growing amongst wheat. It is perhaps better known under the name of "Black-bent." Where it flourishes it proclaims that the land is in a poor condition, for it grows most luxuriantly when the land is in that state.

It appears almost confined to England, as it is rare and local in Scotland. Unknown in Ireland. It is common in the south of Europe, but does not extend north of latitude 56°. It has not been found in America.

In England it is found in Devonshire, Somersetshire, Sussex, Surrey, Kent, Essex, Suffolk, Norfolk, Cambridge, Bedford, Oxford, Leicestershire, Nottinghamshire, Warwickshire, Worcestershire, Cheshire, Yorkshire, Durham, and Northumberland.

Fields and way-sides.

Panicle upright, slender, attenuated, compact, two or three inches long, deposited in short branches all round the rachis. Spikelets oval, consisting of one awned floret of an equal length with the calyx; compressed and numerous. Calyx consisting of two acute membranous glumes of the same length, joined near the base; keels hirsute, and having two green smooth ribs on each side. Floret of one palea, ovate-oblong, and having two green ribs on each side. Awn lengthy, slender, commencing slightly above the base of the palea, and extending considerably beyond it.

Filaments three in number, slender. Anthers protruding, each end notched. Styles united and short. Stigmas long, downy, and two in number. Stem circular, upright, slender, rough, bearing three or four leaves, with rough, striated, swollen sheaths, upper sheath carrying at its apex a blunt downy ligule, and being longer than its leaf. Joints smooth. Leaves flat, rough, striated, and acute. Inflorescence simple, panicled. Length from one to two feet. Root small, fibrous, annual.

This species is easily known by its attenuated panicles, which are often purplish in colour; and by the rough stem and sheaths, and the long dorsal awn.

Dr. Parnell mentions that it can be recognised from *A. pratensis* in the slenderness of the panicle, in the spikelets being larger, the ligules considerably longer, the roughness of the stem and sheaths, and in the keels of the calyx being but slightly hairy; whilst in *A. pratensis* the ligule is short and blunt, keels of calyx and lateral ribs having long hairs, and the stem and sheaths being quite smooth. In *A. geniculatus* the stem and sheaths are also smooth, the awns shorter, the spikelets less, ligule shorter, calyx less acute and different in shape, and the panicle not so tapering. In *A. fulvus* the stem and sheaths are smooth, the panicle less tapering, spikelets less, ligule shorter.

Flowers in the first week in July, and ripens its seeds in October.

My thanks are due to Mr. Joseph Sidebotham, of Manchester, and to Dr. Wilson for good specimens of this Grass.

The illustration is from Dr. Wilson's specimen.

ALOPECURUS BULBOSUS.

LINNÆUS. WILLDENOW. HOOKER AND ARNOTT. SMITH. KNAPP.
BABINGTON. LINDLEY. PARNELL. KUNTH.

PLATE V.—A.

The Bulbous Fox-tail Grass.

Alopecurus—Fox-tail.   *Bulbosus*—Bulbous.

A RARE British Grass, growing in wet salt marshes, and of no use for agricultural purposes.

In England found in the counties of Somerset, Sussex, Gloucester, Suffolk, and Norfolk, most abundant near Yarmouth and Weymouth. In Wales in Cardiff Marshes. Not in Scotland or Ireland.

Abroad it is a native of France, Germany, Spain, Turkey, Greece, Italy, Portugal, and the Mediterranean Islands.

Panicle cylindrical and acuminate. Spikelets numerous, crowded together, consisting of two glumes and one floret. Glumes pointed, equal in length; keels and lateral ribs hairy, separated to the base. Floret slightly shorter than the glumes, consisting of one palea, truncated, on either side with two green ribs. Stigmas long and feathery. Anthers protruding. Styles combined. Stems rising, but bent at the joints, smooth and striated, having three or four leaves with striated smooth sheaths. Joints distant from each other, four in number. Leaves somewhat narrow, smooth underneath, upper surface rough. Inflorescence from an inch to an inch and a half in length. Length from four to fifteen inches. Root tuberous and perennial.

Flowers in July, and ripens its seeds at the end of August.

The present species takes its name from its bulbous root.

It differs from *A. agrestis* in having a smooth stem and sheath, in the truncate summit of the floret, and in the awn extending half its length beyond the palea. In *A. agrestis* the stem and sheaths are rough, floret conical at apex, awn longer.

It differs from *A. pratensis* in having the floret shorter than the glumes, and in being truncate instead of conical at the apex, and in the glumes not being joined at the base.

A. geniculatus is more blunt in the glumes.

A. fulvus has a conical floret with a longer awn.

The illustration is from a specimen gathered in Cheshire by Mr. Joseph Sidebotham, of Manchester.

ALOPECURUS FULVUS.

Smith. Koch. Parnell. Hooker. Lindley.

PLATE V.—B.

Alopecurus geniculatus, var.   Withering.

The Orange Spiked Fox-tail Grass.

Alopecurus—Fox-tail.   *Fulvus*—Fulvous.

THE present species is closely allied to *Alopecurus geniculatus*, but the spike is more slender and not so long. It is found about ponds and ditches, and frequently floating on the water.

It is to be met with in Cheshire, Essex, Worcester, Cambridge, and Norfolk. A rare Scotch Grass, having only been found in Fifeshire and Angusshire. It is not found in Ireland, or America, or Southern Europe, but Linnæus noticed it in Lapland.

Of no use for agricultural purposes. A somewhat rare species.

Panicle upright, compact, cylindrical, having short branches all round the rachis; one to two inches long. Spikelets small, numerous, upright, oval, consisting of one awned floret of the same length as the calyx. Calyx consisting of two equal-sized sharp membranous glumes, three-ribbed; keel fringed, lateral ribs pale green and hairy. Floret of one palea, having two ribs widely apart on each side, oval, slender, a dorsal awn, which does not extend beyond the palea. Filaments three in number, slender. Anthers roundish, yellowish, short, and notched at either extremity. Styles brief and united. Stigmas slender and feathery. Stem ascending, joints bent, base procumbent, smooth, having four or five leaves with smooth striated sheaths, uppermost sheath of same length as its leaf, inflated, and having an oblong ligule, which is membranous. Joints smooth. Leaves acute, flat. Inflorescence panicled. Length from twelve to eighteen inches. Root fibrous and perennial.

Flowers in June.

My thanks are due to Mr. Joseph Sidebotham, of Manchester, for the specimen figured; it was gathered in Rosthern Mere, Cheshire, a locality in which it is not common.

ALOPECURUS GENICULATUS.

LINNÆUS. KOCH. SMITH. LEERS. HOOKER. PARNELL. LINDLEY.
GREVILLE. CURTIS. MARTYN. SCHRADER. SINCLAIR. EHRHART.

PLATE VI.

Alopecurus paniceus,　　　　　　　　OEDER.

The Floating Fox-tail Grass.

Alopecurus—Fox-tail.　　　　　*Geniculatus*—Jointed.

A COMMON Grass, rejected by cattle. Mostly found in moist situations near pools, often floating in the water, yet occasionally to be met with in dry places, where it is more dwarf in habit. When cultivated as an agricultural Grass, the yield is very small.

It is found in Germany, France, Denmark, Norway, Sweden, Lapland, Italy, and in a few parts of the United States. It is not to be found beyond the elevation of two thousand feet.

Panicle upright, one to two inches long, cylindrical, compact, having small branches all round the rachis. Spikelets numerous, ovato, upright, one-awned floret of same length as calyx. Calyx consisting of two equal-sized membranous glumes, blunt, joined at base; lateral ribs hairy; apex purplish. Floret of one palea. Awn slender, extending half its own length beyond the palea. Anthers linear, yellowish. Styles short. Stigmas long and feathery. Stem ascending, joints bent, striated, and polished, uppermost sheath inflated, and of same length as its leaf. Sheaths smooth. Stem bearing branches from the lower joints. Joints dark purple, smooth, long, and narrow. Leaves flat, rough, edges serrated, acute. Inflorescence simple, panicled. Length twelve to fifteen inches. Root fibrous, perennial.

Flowers at the beginning of June, and ripens seed in the last week in August.

The awn in *A. geniculatus* commences slightly above the base, and extends half its length beyond the palea, and the anthers are long;

in *A. fulvus* anthers short, and awn commencing below the centre, and not extending beyond the palea.

A. geniculatus has a smooth stem, whilst in *A. agrestis* it is rough.

In *A. alpinus* panicle short, whilst in *A. geniculatus* it is long.

In *A. pratensis* upper sheath more than twice the length of its leaf, whilst in *A. geniculatus* only of same length as its leaf.

This species is subject to variety.

My thanks are due to Dr. Wilson, of Nottingham, and to Mr. Joseph Sidebotham, of Manchester, for specimens.

The illustration is from Dr. Wilson's specimen, which was gathered in Cheshire.

A

B

PHALARIS CANARIENSIS.

LINNÆUS. SMITH. HOOKER. LINDLEY. PARNELL. KOCH.
GREVILLE. WILLDENOW. KNAPP. MARTYN.
SCHRADER. LEERS. SCHREBER. SINCLAIR. SOWERBY. HUDSON.
WITHERING. SIBTHORP.

PLATE VII.—A.

The Cultivated Canary Grass.

Phalaris—Shining (from the Greek).   *Canariensis*—Canary Island.

PHALARIS. *Linnæus.*—Panicle spiked or spreading, with laterally compressed spikelets. Two glumes nearly equal in size, upright, membranaceous. Glumellas two, awnless, hairy; outer palea without lateral ribs. Leaves broad and flat. Only two British species.

Its name is derived from the Greek, in allusion to the polished appearance of its seeds.

THIS handsome Grass is not strictly a native of Great Britain, yet has now become naturalized in many parts of England and Scotland, probably owing to the extensive use of its seeds for feeding Canaries and other small birds.

It appears to prefer rich ground, and near Boeston it is seldom found except in gardens and orchards.

Native of the Canary Islands. It has also become naturalized in America.

Probably it is not of any agricultural value.

Panicle globular, upright, with brief branches. Spikelets oval, imbricated, flat, handsomely marked with yellowish green and white stripes, having one awnless floret. Calyx of two equal-sized compressed glumes. Floret consisting of two paleæ, the outer one egg-shaped, acute, hairy, having two membranous lance-shaped acute scales at the base, of half the length of the palea. Palea having the outer one longest. Seeds polished. Stem slender, upright, smooth, having five or six leaves with rough inflated sheaths, the upper one being longer than its leaf, and having a white rounded ligule at its apex. Joints yellowish, naked. Leaves somewhat broad, lanceolate, acute, rather rough. Inflorescence panicled. Length from twelve to twenty-four inches. Root fibrous, white, annual.

Flowers at the beginning of July, and ripens seed in last week in August.

Specimens have been forwarded by Dr. Wilson, of Nottingham, and Mr. Sidebotham, of Manchester, the latter from the road-sides in Cheshire, where it is not common.

The illustration is from a specimen gathered in an orchard at Beeston, near Nottingham, where it grows sparingly.

PHALARIS ARUNDINACEA.

LINNÆUS. SMITH. HOOKER. KOCH. GREVILLE. PARNELL. HUDSON.
PURTON. SCHRADER. OEDER. LEERS. EHRHART.

PLATE VII.—B.

Arundo colorata,	SOLAND. DRYANDER. SMITH.
" "	KNAPP. WILLDENOW. HALLER.
Phalaris arenaria,	SMITH. SOWERBY. HUDSON.
" *phleoides,* var.	AITON.
Phleum arenarium,	LINNÆUS. WITHERING.

The Reed Canary Grass.

Phalaris—Shining (from the Greek).   *Arundinacea*—A reed.

A HANDSOME and abundant coarse-growing species on the banks of rivers and sides of lakes, preferring a strong clayey soil. Cattle are not partial to it, yet it produces a large and early crop, and may be cut three times a year. A variety cultivated in our gardens is exceedingly handsome; it is best known as the "Ribbon Grass," or "Painted Lady Grass." The leaves are beautifully striped with green and white, varying considerably in the width of the different bands of colour.

Common in Scotland, England, Ireland, Germany, and in the South of Europe, but unknown in America, Lapland, Sweden, or Norway.

Panicle upright, long, and narrow; rachis and branches rough. Spikelets crowded, numerous, of one awnless floret hid within the calyx. Spikelets occasionally tinged with purple, white, yellow, and green. Calyx two nearly equal, acute glumes; keels toothed, sides rough; calyx three-ribbed. Floret of two paleæ, the outer one acute, rough; edges hairy, longer than inner palea. Stem upright, circular, smooth, having five or six leaves with sheaths that are smooth and striated; upper sheath considerably longer than its leaf, having a lengthy decurrent membranous ligule at its apex; the other ligules more blunt. Joints smooth and purple. Leaves broad, pale green, acute, flat, ribbed, the central rib bolder than the others, rough, edges toothed minutely. Inflorescence compound panicled. A tall Grass, growing from two to five feet in height. Roots perennial, creeping horizontally.

Flowers in second week of July, and the seeds become ripe in the middle of August.

My thanks are due to Dr. Wilson, of Nottingham, and to Mr. Joseph Sidebotham, of Manchester, for specimens of this species.

The illustration is from Mr. Sidebotham's specimen.

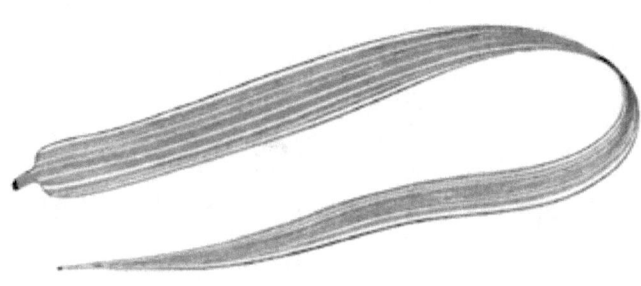

Ribbon-Grass.

AMMOPHILA ARUNDINACEA.

HOST. HOOKER. PARNELL.

PLATE VIII.—A.

Arundo arenaria,	SMITH. HOOKER. GREVILLE.
" "	LINNÆUS. WILLDENOW. KNAPP.
" "	MARTYN. DICKSON. SCHRADER.
" "	OEDER. EHRHART.
Ammophila arenaria,	LINDLEY. KOCH.
Calamagrostis arenaria,	ROTH. WITHERING.
Spartum anglicanum,	GERARDE.

The Sea Reed.

Ammophila—To love the sand (from the Greek).   *Arundinacea*—A reed.

AMMOPHILA. *Host.*—Only one British representative of this genus. Panicle spiked, with laterally compressed spikelets. Nearly equal-sized keeled membranous glumes, longer than the floret. Glumellas two, hairy at the base. Outer palea five-ribbed. Leaves narrow.

It takes its name from two Greek words alluding to its habitat—sand near the sea shore.

A HANDSOME very coarse Grass, of no agricultural value, as no cattle will eat it, yet valuable as growing amongst sand near the sea, and thus preventing by its matted creeping roots that inroad of the sea which would otherwise take place. It is known as the Common Sea Grass, the Marum or Matweed.

The present species is protected by an Act of Parliament, on account of its great use along our coasts in Norfolk, and in Holland it is extensively grown, as also about Lytham, in Lancashire.

Found on coasts of Northumberland, Durham, Lancashire, Cheshire, Denbigh, Anglesea, Merioneth, Essex, Kent, Norfolk, Somersetshire, Worcester, Devonshire, and Cornwall. Common in Scotland and in the Orkney Islands, and is a native of Sweden, Lapland, Norway, North America, United States, and in the Mediterranean Islands.

Panicle dense, upright, lengthened-oval, short rough branches; panicle three to five inches in length. Spikelets numerous, narrow, long, one floret, which is shorter than the calyx. Calyx consisting of two unequal-sized acute glumes, devoid of lateral ribs. Floret of two paleæ, the outer one five-ribbed, the dorsal rib toothed; base slightly hairy. Palea equal in length. Stem upright, smooth, polished, circular, carrying three or four leaves with somewhat rough sheaths, uppermost sheath of about the same length as its leaf, having a lengthened lance-shaped membranous ligule at its apex. Joints smooth. Leaves narrow, smooth, pointed, glaucous, and rigid. Inflorescence panicled. Length from eighteen to twenty-four inches. Root creeping and perennial.

Flowers at the beginning of July.

The illustration is from a specimen procured at Lytham.

PHLEUM PRATENSE.

LINNÆUS. KOCH. LEERS. PARNELL. SMITH. HOOKER. WILLDENOW.
LINDLEY. GREVILLE. KNAPP.
MARTYN. SCHREBER. SCHRADER. SINCLAIR. SOWERBY. HUDSON.
WITHERING. SIBTHORP. ABBOT. RELH. HULL.

PLATE VIII.—B.

Phleum nodosum,	LINNÆUS. WILLDENOW. LEERS.
" "	OEDER. SINCLAIR.
Alopecurus bulbosus,	DICKSON.
Gramen typhoides minus,	MORISON.
" typhinum minus,	GERARDE.

The Cat's-tail Grass.

Phleum—Reed Mace.   *Pratense*—Meadow.

PHLEUM. *Linnæus.*—Stigmas long and slender. Floret consisting of two paleæ, which are not awned. Panicle spiked. Spikelets laterally compressed. Glumes parallel, about equal in size, longer than the floret. There are six British species.

Name derived from the Greek, formerly applied, it is conjectured, to the *Reed Mace.*

A COMMON species, known as Timothy Grass, growing in moist soils, common in Great Britain in meadows and pastures. It has been recommended as a good agricultural Grass, from the nutritive matter in its flower stems, yet cattle are not fond of it, and it appears to be of too slow growth for the aftermath, in order for it to become a remunerative Grass. Mr. Sinclair condemns its being grown alone in fields, but recommends it amongst other species as a valuable addition.

Found in Lapland, Sweden, Norway, and North America.

Panicle upright, compact, and cylindrical, green, and white; length varying from two to five inches. Spikelets diminutive, abundant, arranged in pairs all round the rachis on brief foot-stalks, having one slightly-awned floret considerably shorter than the calyx. Calyx consisting of two equal-sized glumes, having a broad membranous margin, keels furnished with short stout white hairs. Palea, outer one five-ribbed, egg-shaped, apex jagged, keel hairy. Floret consisting of two paleæ, which are membranous. Stem circular, smooth, upright, bearing four or five leaves. Sheaths smooth, the uppermost one longer than its leaf, having a membranous ligule. Joints smooth. Leaves rough, flat, broadish, acute. Inflorescence simple, panicled. Length eighteen to twenty-four inches. Root creeping, somewhat bulbous, perennial.

Flowers in the third week in June, and ripens seed at the end of July.

In *A. alpinus* the glumes are a third longer than their awns, in *P. pratense* about twice the length.

In *P. arenarium* glumes acute and not awned, in *P. pratense* blunt and awned. In *P. arenarium* floret one third of the length of the calyx, whilst in *P. pratense* about half the length of the calyx.

P. michelii has longer spikelets, acute glumes, and not awned. *P. pratense*, var. *longiaristatum*, Parnell, (The Long-awned Timothy Grass,) found in a damp wood near Edinburgh, differs from the normal form by the awns of the glumes being almost as long as the glumes, and the root being bulbous. It does not flower till August.

P. pratense, var. *longiciliatum*, Parnell, (The Bulbous Timothy Grass.) Stem near base prostrate, joints bent, awns of glumes short, root bulbous. Found in sandy situations. Flowering in the end of July.

Fine specimens have been forwarded both by Dr. Wilson, of Nottingham, and Mr. Sidebotham, of Manchester.

The illustration is from Dr. Wilson's specimen.

PHLEUM ALPINUM.

Linnæus. Willdenow. J. E. Smith. Dickson. Don. Hooker.
Schrader. Oeder. Withering. Koch. Parnell.

PLATE IX.—A.

Phleum commutatum,   Gaudichaud.

The Alpine Cat's-tail Grass.

Phleum— Reed Mace.   *Alpinum*—Alpine.

EXCLUSIVELY a mountain Grass, flourishing in damp situations, at elevations varying from two thousand five hundred to three thousand five hundred feet. It is by no means a common species, being unknown either in England or Ireland. Mr. Dickson, the discoverer, first recognised *Phleum alpinum* near Garway Moor, whilst other explorers have found it on Craigneulict, above Killin, Ben Lawers, Clova Mountains, Breadalbane Mountains, Feula Burn, Canlochen Glen, Glashieburn, Glen Fiadh, and Loch Brandy.

Abroad it is plentiful in Norway, Sweden, Lapland, Switzerland, Germany, and North America.

As an agricultural Grass it is useless.

Root slightly creeping, perennial, and somewhat tuberous. Stem upright, except near the base, where it is usually decumbent, smooth, circular in form, with four or five leaves, which are provided with polished striated sheaths. Uppermost sheath slightly inflated, and being longer than its leaf; capped with a short obtuse ligule. Joints smooth. Leaves broad, sharp pointed, smooth both above and below, except along the edges, where rough. Inflorescence simple, panicled. Panicle oval, tinged green and white with dull purple, varying in length from half an inch to an inch and a half. Spikelets close together, diminutive and numerous. Calyx of two equal-length glumes, with wide membranous margins; the keels, which are green, are fringed with short white hairs; the glumes end in a rough awn, of rather more than half the length of the glume. Floret consisting of a pair of membranous paleæ, the exterior one being egg-shaped, five-ribbed; keel hirsute; small rough dorsal awn; inner palea somewhat shorter. Length from six to twelve inches.

Comes into bloom in July, and the seed becomes ripe before September.

The present Grass bears some resemblance to *Alopecurus alpinus*, yet the latter species has the panicle silky, the glumes of the calyx destitute of awns, and the floret possessing only one palea.

I am indebted to Mr. Joseph Sidebotham, of Manchester, for the specimen illustrated, which was gathered on the Clova Mountains.

PHLEUM ASPERUM.

JACQUIN. KOCH. SMITH. HOOKER. BABINGTON. LINDLEY.
SCHRADER. PARNELL. VILLARS.

PLATE IX.—B.

Phleum paniculatum,	HUDSON. SMITH. KNAPP. AITON.
" *viride,*	ALLIONI.
Phalaris aspera,	RETZIUS. WILLDENOW. HOST.
" *paniculata,*	AITON. SIBTHORP.

The Rough Cat's-tail Grass.

Phleum—Reed Mace.   *Asperum*—Rough.

A RARE useless agricultural Grass, limited to the western portion of Great Britain, being confined to the counties of Gloucester, Bedford, Oxford, Cambridge, and Norfolk, chiefly in the two last-mentioned counties. Mr. Hudson procured it near Bristol, and on the heath at Newmarket; Mr. Crowe near Bournbridge. It has also been seen in Badminton Park. In Ireland it has been found near Belfast.

It is a native of Belgium, Italy, Switzerland, Holland, Prussia, and France.

Phleum asperum grows in arid sandy situations.

Root perennial, consisting of a number of strong fibres. Stem circular, upright, exceedingly smooth, carrying four or five flat, rough, acute leaves, with rough tumid sheaths, the upper one extending beyond its leaf. Ligule bold and pointed. Joints four in number, covered by the sheaths. Inflorescence close, panicled, and from two to five inches in length. Spikelets abundant, compact, composed of two equal length, rough, cuneate glumes, and one floret. The glumes variegated with green and white, the inner edge obtuse at the apex, straight and membranous. The floret consisting of two paleæ, the exterior one rough, indistinctly five-ribbed, the centre rib being hirsute on the upper portion. Floret a third shorter in length than the glumes. Filaments capillary, three in number. Anthers cloven at either extremity. Styles bold, two in number. Stigmas feathery. Seed diminutive, loose, cylindrical in form. Length from six to eighteen inches.

Flowers in July, the seed becoming ripe at the commencement of September.

The present species is readily distinguishable from the remainder of this family by its cuneate glumes, and rough but not hairy keels.

The specimen illustrated was gathered near Belfast, and contributed by Mr. Joseph Sidebotham, of Manchester.

PHLEUM MICHELII.

ALLIONI.   KOCH.   SMITH.   HOOKER.   LINDLEY.   PARNELL.   SCHRADER.

Phalaris alpina,   HŒNKE.   HOST.

The Michelian Cat's-tail Grass.

Phleum—Reed Mace.   *Michelii*—After Micheli.

THE present species is a south of Europe Grass, which has been added to our British flora, from the circumstance that Mr. Don discovered it amongst the rocks on the higher parts of the Clova Mountains, in Scotland. It grows from one to two feet high.

Phleum michelii differs from *P. arenarium* in having perennial roots, in the whole of the keels of the glumes being hirsute, and in the floret being entire at the apex instead of being notched. It is also readily distinguished from *P. pratense,* in having the glumes of the calyx acute-lanceolate, instead of being obtuse; and from *P. boehmeri* by the tapering glumes.

No other botanist has been fortunate enough to discover it.

It must be considered a doubtful British species.

PHLEUM BOEHMERI.

SCHRADER. KOCH. SMITH. KUNTH. HOOKER. LINDLEY. WITHERING. BABINGTON. PARNELL.

PLATE X.—A.

Phalaris phleoides,	LINNÆUS. WILLDENOW. SMITH.
" "	OEDER. HOST. EHRHART.
" "	SINCLAIR.
Chilochoa boehmeri,	BEAUVOIS.

The Purple-stalked Cat's-tail Grass.

Phleum—Reed Mace.   *Boehmeri*—After Boehmer.

A RARE British Grass, of no agricultural use; almost confined to chalky or dry sandy fields in Cambridgeshire and Norfolk, in the latter county near Narburgh. It has not been discovered either in Scotland or Ireland.

On the continent it is included in the flora of Russia, Switzerland, Italy, France, Germany, Norway, and Sweden.

Root fibrous and perennial. Stem upright, smooth, simple, slender, striated; upper portion purple and shining, having four or five leaves, with smooth rather tumid striated sheaths, the uppermost one much longer than its leaf, and having a broad obtuse ligule, which entirely encloses the stem, mostly four-jointed, the joints being all below the centre. Leaves rough on both sides and along the edges, flat, linear-lanceolate in form, except those near the base, which are narrower. Inflorescence compact, dense, panicled, cylindrical. Spikelets diminutive, very numerous, situated all round the panicle, consisting of two equal-sized glumes and one floret; glumes linear, the edges being white and membranous; apex oblique; floret awnless, and only three fourths of the length of the glumes, consisting of two equal-sized paleæ, the exterior one being five-ribbed and roughish, the inner one membranous.

Ovarium hirsute, scales bold, styles two, stigmas feathery, stamens three. Length from six to eighteen inches.

Flowers in July, and the seeds become ripe about the middle of August.

The present species differs from *Phleum asperum* in not having the glumes cuneate, in not having a long and pointed ligule, and in having the keels fringed. It differs from *P. arenarium* in having the floret entire instead of jagged at the apex, in the floret being much longer in comparison to the glumes, and in the inner edges of the glumes not being fringed with diminutive hairs as in *P. arenarium*. From *P. michelii* in having more linear-shaped glumes, and in having only the upper portions of the keels hirsute, instead of throughout their whole length, as in *P. michelii*; whilst from *P. pratense* it differs in the glumes being pointed and destitute of awns, and the inner edges ending obliquely instead of abruptly. Also in the exterior palea having the apex entire instead of jagged, as in *P. pratense*.

The specimen illustrated was procured in Norfolk by Mr. Joseph Sidebotham, of Manchester.

PHLEUM ARENARIUM.

LINNÆUS. SMITH. HOOKER. LINDLEY. KOCH. GREVILLE. PARNELL.
SCHRADER. OEDER. EHRHART.

PLATE X.—B.

Phalaris arenaria.   KNAPP. HUDSON. WILLDENOW.

The Sea Cat's-tail Grass.

Phleum—Reed Mace.   *Arenarium*—Sea-shore.

THE *Phleum arenarium* is almost exclusively a sea-side Grass, growing in loose sand. It is to be met with on the coasts of Devonshire, Somersetshire, Sussex, Kent, Suffolk, Norfolk, Cheshire, Durham, and Northumberland. Also in Denbigh and Fifeshire; indeed it is by no means uncommon in Scotland, although local in Ireland. On the continent it is met with in various places in southern Europe. Inland it is recorded as growing on Swaffham and Newmarket Heaths.

The leaves being harsh it is not an agricultural Grass.

Root annual, consisting of numerous long simple fibres. Stem circular, smooth, and mostly having a purple tinge on the upper portion; joints naked. The stem bears four or five leaves, whose sheaths are slightly tumid, smooth, and striated, the uppermost sheath being above double the length of its leaf. Leaves rough both above and below, brief and broad. Inflorescence simple panicled, the panicle being obovate-cylindrical in form, and upright in habit. Spikelets oval in shape, and numerous, consisting of one floret of one third the length of the calyx, and awnless. Calyx composed of a couple of equal-sized membranous glumes, which are lanceolate in form. Upper portion of the keel and inner edges fringed. Floret consisting of two equal-sized membranous paleæ, notched at the apex, the outer paleæ five-ribbed, keel hirsute. The length of this Grass varies considerably, according to the support it is enabled to procure from the sandy ground.

Sometimes it does not exceed three inches, at others it is five times this length.

Phleum arenarium has much smaller spikelets than *P. michelli*, it is notched at the summit instead of being entire, and the glumes not hirsute on the lower half of their keels. It differs from *P. pratense* in being smaller, the base of the panicle contracted, and having no awned floret.

Comes into flower in the middle of July, the seeds becoming ripe in about a month.

The specimen illustrated was gathered at Fleetwood, by Mr. Joseph Sidebotham, of Manchester.

Phleum arenarium.—From a dwarf specimen.

LACURUS OVATUS.   CASTRIDIUM LENDIGERUM.

LAGURUS OVATUS.

LINNÆUS. HOOKER AND ARNOTT. KOCH. KUNTH. SMITH. BABINGTON.
LINDLEY. KNAPP. WITHERING. PARNELL.
WILLDENOW. DICKSON. SCHRADER. HOOKER. SCHREBER. HULL.

PLATE XI.—A.

Alopecuros genuina,   MORISON.
" *spica rotundiore,*   MORISON.

The Hare's-tail Grass.

Lagurus—Hare's-tail (from the Greek).   *Ovatus*—Egg-shaped.

LAGURUS. *Linnæus.*—Panicle spiked. Spikelets laterally compressed. Glumes fringed throughout, terminating in a lengthy subulate point. Glumellas two in number, membranaceous in texture, the exterior one terminating in two long bristles. Only one British example, and this confined to a portion of Guernsey. The name is derived from the Greek, and signifies a hare's tail, from the downy feel and appearance of the panicle.

ONE of the rarest and most beautiful of our British Grasses, growing in sandy exposed situations in the north and west of Guernsey, one of the Channel Islands. It was first discovered there by Mr. Gosselin. Miss Guille informs me that it is abundant near the seashore. Sir J. E. Smith, in his "British Flora," remarks that *Lagurus ovatus* serves to decorate flower-pots in winter, like the *Stipa pennata,* and the foreign *Briza maxima;* there are, however, a number of other species which, when placed in a vase in a bunch, produce a pleasing effect.

Root annual, composed of seven or eight woolly fibres. Stem upright, circular, smooth, with three or four joints. Leaves four or five in number, with tumid, very downy sheaths. Ligule bold, obtuse, and encircling the stem. Leaves flat, lanceolate in shape, short, pointed, densely covered both above and below with downy hairs. Inflorescence compound panicled. Form ovate, except near the base, where it is more flat. An inch in length. Branches short and crowded; at first upright, but afterwards driven to one side by the power of the wind over them. Spikelets dense, consisting of two equal-sized long glumes, fringed with long, white, downy hairs, and one floret, which is shorter than the glumes, and composed of two equal sized paleæ, the exterior one being rough and five-ribbed; ending in two bristles, and having a lengthy dorsal awn. Styles two. Filaments three. Stigmas feathery. Anthers cloven at either extremity. Length from three to nine inches.

Flowers in June, and the seed becomes ripe at the end of July.

I am indebted to Miss Guille for the specimen illustrated.

GASTRIDIUM LENDIGERUM.

BEAUVOIS. HOOKER AND ARNOTT. PARNELL. GAUDICHAUD. LINDLEY. LINK. BABINGTON.

PLATE XI.—B.

Gastridium australe,	BEAUVOIS. KUNTH.
Milium lendigerum,	LINNÆUS. SMITH. WILLDENOW.
" "	SCHREBER. HULL.
Agrostis ventricosa,	KNAPP. GOUAN.
" *australis*,	LINNÆUS.
" *rubra*,	HUDSON.
Alopecurus ventricosus,	HUDSON.

The Nit Grass.

Gastridium—A swelling.   *Lendigerum*—Maggot-bearing.

GASTRIDIUM. *Beauvois.*—A solitary Grass of this genus belongs to Great Britain, the *Gastridium lendigerum*; having a spiked contracted panicle, with two ventricose, acute, awnless glumes, which are membranaceous, keeled upwards, and considerably longer than the floret. Glumellas two, also membranaceous. Named from the Greek, in allusion to a little swelling which occurs at the base of the spikelet.

A SOMEWHAT rare species, having a glossy swollen appearance at the base of the glumes. Found in open fields where water has stagnated, and more especially near the sea.

Found in **Norfolk, Essex, Kent, Sussex, Surrey, Devonshire, Dorsetshire, Hampshire, Somersetshire, Gloucestershire, Denbigh,** and **Flint. Isle of Wight,** abundant. **Isle of Sheppy.**

Abroad it is a native of **France, Germany, Italy, Switzerland, Spain, Portugal, Turkey, Greece, Northern Africa,** and the **Mediterranean Islands.**

It has not been found either in Scotland or Ireland.

Of no use as an agricultural Grass.

Root annual and fibrous, having slender branching fibres. Stem upright, circular, polished; carrying four or five flat, acute, rough leaves, with usually smooth sheaths, the upper one being longer than its leaf. Joints mostly three in number. Ligule conspicuous, pointed, and broad. Inflorescence compound panicled, compact; pale green in colour; branches rough. Rachis circular and smooth. Spikelets upright, numerous, composed of two unequal-sized acute glumes, that are tumid at the base; deeply dentate at the upper portion, and with green keels, and one floret, two thirds less in length than the glumes, and consisting of two paleæ, the exterior one being five-ribbed, with the summit jagged; inner paleæ somewhat shorter, with smooth lateral ribs. Awn rough, yet slender, twice the length of the paleæ. Styles brief, distant, and two in number. Stigmas feathery. Filaments three, slender. Anthers notched at either extremity. Scales acute.

Gastridium lendigerum does not flower till August, nor ripen its seeds before the end of September.

The specimen illustrated was gathered on St. Vincent Rocks, Bristol, by Mr. Joseph Sidebotham, of Manchester.

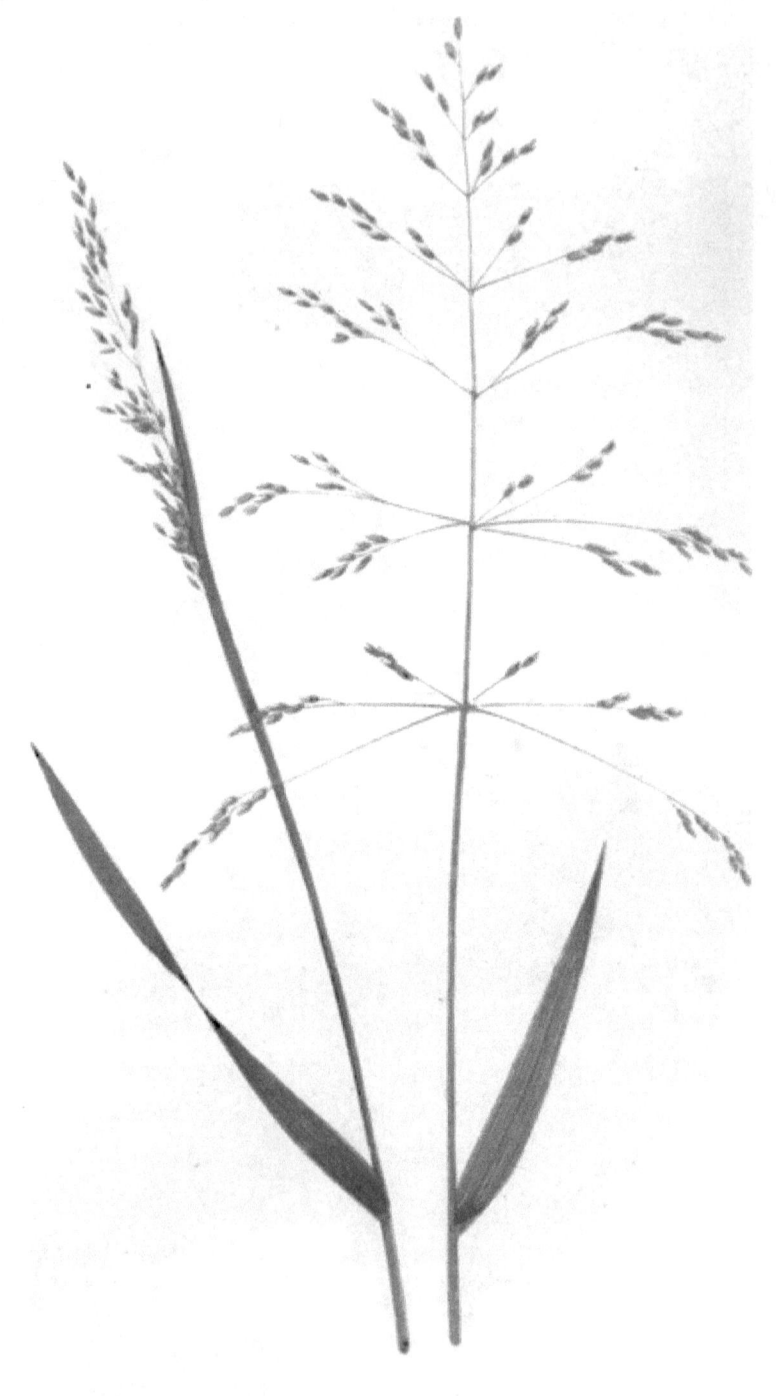

MILIUM EFFUSUM.

Linnæus. Smith. Parnell. Hooker and Arnott. Lindley. Greville. Koch. Hudson. Withering. Hull. Relhan. Sibthorp. Abbot. Curtis. Knapp. Leers. Schrader. Sinclair.

PLATE XII.

Gramen miliaceum,   Ray. Gerarde.
"   "   *vulgare,*   Morison.

The Spreading Millet Grass.

Milium—Millet.   *Effusum*—Spreading.

Milium. *Linnæus.*—Confined to one British species, with spreading panicle, having in some degree dorsally compressed awnless spikelets, with two almost equal-sized glumes, and the same number of glumellas. Millet Grass, says Sir W. Hooker, either receives its name from *mille*—a thousand, on account of its fertility, or from *mil*—a stone, because of the hardness of its seeds.

AN elegant Grass, growing commonly in damp woods and in moist shady situations. Of no agricultural value, but the seeds are a favourite food of small birds.

Common in many portions of England, Scotland, and Ireland.

Abroad it is found in Norway, Sweden, Lapland, North America, the United States, and in the Mediterranean Islands.

Root fibrous, perennial, and branching. Stem upright, smooth, slender, shining; and having four or five broad, flat, pale green, shining, smooth, lanceolate-shaped leaves, with sheaths that are smooth and striated, the upper sheath having an oblong membranous ligule. Joints smooth. Inflorescence compound panicled, or spreading, the panicle being glabrous, subverticillate, loose, and of large size, with lengthy slender branches arranged in alternate distant clusters along the rachis. Spikelets abundant, small, ovate in form, on delicate footstalks, and consisting of one awnless floret, hid within the calyx. The calyx composed of two equal sized, rough, three-ribbed, broad, membranous glumes. The floret composed of two equal-sized membranous paleæ. Styles short. Stigmas feathery. Length from three to four feet.

Flowers about the middle of June, and ripens its seeds about the middle of August. Colour pale whitish green.

For specimens from Reddish Woods, Cheshire, I am indebted to Mr. Joseph Sidebotham, of Manchester; and for others from Falmouth, to Mr. H. C. Bastian, of Falmouth.

The specimen illustrated was gathered in Reddish Woods, Cheshire, by Mr. J. Sidebotham, of Manchester.

STIPA PINNATA.
XIII

STIPA PENNATA.

Linnæus. Hooker and Arnott. Smith. Parnell. Hudson.
Withering. Hull. Willdenow. Knapp. Sinclair. Schrader.
Koch. Lindley. Babington.

PLATE XIII.

Spartum austriacum,   Gerarde.
Gramen sparteum pennatum,   Bauhin. Dillenius.

The Common Feather-Grass.

Stipa—Tow or flax (from the Greek).   *Pennata. Penna*—A feather.

Stipa. *Linnæus.*—Panicle upright and contracted. Glumes membranaceous, two in number, longer than the floret. Floret stipulate. Glumellas cartilaginous; exterior glumella involute, ending in a twisted awn of great length, which is jointed at the base. There is only a solitary representative of this genus in Great Britain, and that one very rare. The name is derived from the Greek, signifying a flaxen appearance, in reference to the present species—*Stipa pennata.*

AN exceedingly rare Grass, of but little agricultural value. It was first discovered on rocks in Long Sleadale, Westmoreland, by Dr. Richardson and Mr. Lawson, in the time of Dillenius, and from that time there seems to be no re-discovery of it. Lately, however, Mr. Joseph Sidebotham has forwarded me specimens gathered in corn-fields near Hebden-Bridge, Yorkshire, where he informs me it is found growing wild, although rare.

From the extreme beauty of this species it is extensively cultivated in our gardens, and its flowers gathered to ornament our drawing-rooms during winter.

Stipa pennata grows in dry sandy situations. It is found in many parts of Germany.

Stem upright, circular, smooth, carrying four or five long, splendid, hirsute, rigid, and setaceous leaves, with rather rough sheaths, the uppermost one being longer than its leaf. Joints three or four in number, hid by the sheaths. Ligule of uppermost sheath hirsute, long and pointed. Inflorescence racemed, rising out of the highest sheath. Spikelets consisting of two nearly equal-sized, long, slender, hirsute glumes; and one floret of only half the length of the glumes, consisting of two paleæ, the exterior one sharp-pointed at the base, hirsute, five-ribbed, and ending in a considerable feather-like awn. The awn commences at the apex of the exterior palea, and usually twenty times its length; with the exception of that portion immediately about the base, it is feathery to the apex. Styles prominent, two. Stigmas feathery. Filaments capillary, three in number. Anthers notched at either extremity. Scales acute. Length about two feet. Root perennial and fibrous.

This species does not flower till August, and its seeds become ripe about the middle of September.

When gathered to decorate a room, this should be done at the commencement of September, before the seeds are ripe.

The illustration is from a specimen gathered near Hebden Bridge, Yorkshire, by the late S. Gibson, and forwarded by Mr. Joseph Sidebotham, of Manchester.

POLYPOGON MONSPELIENSIS.   P. LITTORALIS.
XIV

POLYPOGON MONSPELIENSIS.

DESFONTAINES. KOCH. SMITH. HOOKER. LINDLEY.
PARNELL. SCHRADER. MARSCHALL.

PLATE XIV.—A.

Agrostis panicea,	AITON. WILLDENOW.
Alopecurus monspeliensis,	LINNÆUS. WITHERING.
" *aristatus,*	HUDSON.
Agrostis triaristata,	KNAPP.
Alopecurus maxima anglica,	RAY.
" *paniceus,*	LINNÆUS. WITHERING.
Cynosurus paniceus,	LINNÆUS.
Phleum crinitum,	SCHREBER. J. E. SMITH.

The Annual Beard-Grass.

Polypogon—Many—A beard (from the Greek). *Monspeliensis*—Belonging to Montpelier.

POLYPOGON. *Desfontaines.*—The Beard-Grass, of which two species occur in Great Britain, has compact panicles, whose spikelets are laterally compressed. Glumes two, equal-sized, notched, and awned. Name derived from the Greek, meaning *many beards.*

A BEAUTIFUL rare Grass, growing in moist situations near the sea. Found in Hampshire, Kent, Essex, Norfolk, Gloucester, Durham, Fifeshire, and the Island of Guernsey. Also found along the Mediterranean Sea.

Stem upright, circular, rather rough, carrying five or six broad flat acute rough leaves, with striated smooth sheaths, the upper one extending beyond its leaf. Ligule long, rough, and acute. Joints smooth. Inflorescence compound panicled. Panicle upright, close, lobed, silky; length from one to two inches. Branches rough, but rachis almost smooth. Spikelet of one awned floret, shorter than the calyx. The calyx of two membranous hairy obtuse linear glumes, having a rough slender awn of great length arising just beneath the apex. Dentate on the lower half of the keels. Destitute of lateral ribs. Floret consisting of two paleæ, the exterior one ovate in shape, destitute of lateral ribs, half the length of the calyx, and furnished with a small awn, the interior one thin and pellucid, somewhat shorter and with entire margins. Length from nine to fifteen inches. Root creeping and fibrous.

Flowers in the first week in July, and ripens its seed the second week of August.

The illustration is from a specimen gathered in Plumstead Marsh, Kent, and forwarded by Mr. Joseph Sidebotham, of Manchester.

POLYPOGON LITTORALIS.

SMITH. HOOKER. LINDLEY. BABINGTON. PARNELL. KUNTH. KOCH.

PLATE XIV.—B.

Agrostis littoralis,   J. E. SMITH. WITHERING.
"   "   KNAPP. DICKSON.

The Perennial Beard-Grass.

Polypogon—Many—A beard (from the Greek).   *Littoralis*—Sea-shore.

A VERY rare species, growing in salt marshes. Found in Norfolk, near Cley; in Essex, on the coast; Hampshire, near Porchester; and Kent, near the Woolwich powder-magazine; and in Germany.

Stem upright, circular, smooth, carrying seven or eight flat, roughish, acute leaves, with striated yet smooth sheaths, the uppermost one considerably longer than its leaf, and its ligule bold, acute, and about twice as long as it is broad. Joints smooth. Inflorescence compound panicled, the rachis and branches being rough with minute teeth. Spikelets numerous, laterally compressed, small, and composed of two equal-sized, linear, obtuse, hirsute glumes, and one floret of a little above half the length of the glumes. Glumes destitute of lateral ribs, dentate on the keel, and having a long rough awn of the same length as the glumes, arising just beneath the apex. Floret consisting of two paleæ, the exterior one destitute of lateral ribs, having a slender awn commencing slightly beneath the apex. Inner palea shorter, thin, pellucid, and having entire margins. Stamens two; styles two; scales two. Stigmas feathery. Length from six to twelve inches. Root perennial, and somewhat creeping.

In *Polypogon monspeliensis* the awns of the glumes are above twice the length of the glumes.

The illustration is from a specimen found near the powder-magazine, Kent, forwarded by Mr. J. Sidebotham, of Manchester.

CALAMAGROSTIS EPIGEJOS.

ROTH. LINDLEY. KOCH. HOOKER. PARNELL.

PLATE XV.—A.

Arundo epigejos,	SMITH. LINNÆUS. WILLDENOW.
" "	KNAPP. SCHRADER. EHRHART.
" *calamagrostis*,	HOOKER. LIGHTFOOT. HUDSON.
Calamagrostis lanceolata,	WITHERING. (*not of* ROTH.)

The Wood Reed.

Calamagrostis—Palm Grass. *Epigejos*—Upon—The earth
(from the Greek).*

CALAMAGROSTIS. *Adanson.*—Of *Calamagrostis* there are three British species, all growing in moist situations. In *C. epigejos* and *C. stricta* the panicle is upright and close; in *C. lanceolata* it is loose. Spikelets laterally compressed. Two almost equal-sized glumes. Two membranaceous glumellas. Name derived from the Greek, signifying the Palm Grass, a very inappropriate name.

THE *Calamagrostis epigejos* is by no means a common Grass, growing in damp woods and shady ditches.

It is to be met with in Dalrymple Wood, Ayr; and in Argyle and

---

* This should be spelt and pronounced *Epigeios* with the g hard. The term is intended to characterize the species as one that grows on the *land*, as distinguished from the sea or from watery places. Linnæus gives as its habitat "*Dry* hills, banks, and corners of fields." Withering, Hudson, Babington, and other botanists speak of "moist shady places, wet hedges," etc. Two localities in which it was found near Nottingham, by Dr. Howitt and Dr. Wilson, are *dry*. Doubtless it does grow in *moist* places, but differs from the other Reed Grasses in not being *limited* to such places as they are. Hence the propriety of the specific term "*Epigeios*."

Aberdeenshire, Scotland; and near St. Ann's Wells, Nottingham, although rare. The last locality is from the authority of Mr. J. Sidebotham. Also in Somerset, Dorset, Sussex, Kent Surrey, Middlesex, Essex, Norfolk, Suffolk, Cambridge, Bedford, Oxford, Leicester, Warwick, Worcester, Shropshire, Lincoln, York, Cumberland, Durham, Northumberland, and Anglesea. Abroad in Germany, Denmark, Sweden, Norway, and Lapland.

Stem circular, upright, and somewhat rough, carrying four narrow, acute, taper-pointed leaves, with inner surface and edges rough, but smooth on the back. Sheaths smooth, striated, the uppermost one extending beyond its leaf, and having at its apex a lengthy, lanceolate-shaped, divided ligule. Joints smooth. Inflorescence brown, compound panicled. Panicle upright, compact, four inches in length. Branches and rachis rough, the branches in alternate clusters. Calyx composed of two narrow, acute, equal-sized glumes, destitute of lateral ribs; containing one awned floret, not so long as the glumes. Floret consisting of two paleæ, the exterior one ovate-lanceolate, destitute of lateral ribs, and ending in two bifid points; at the base a number of long straight hairs, colourless, and equal in length to the calyx. Awn long and slender, commencing at the centre of the palea, and rising to the summit of the hairs. Inner palea acute, membranous, linear, and considerably shorter. Length from three to five feet. Root perennial and creeping.

Flowers at the end of July, and seed ripens at the end of August. Cattle will seldom eat this Grass.

Distinguished from *C. stricta* in the hairs and awn of *C. stricta* scarcely extending beyond the floret.

The illustration is from a specimen gathered near St. Ann's Wells, Nottingham.

CALAMAGROSTIS LANCEOLATA.

Roth. Koch. Kunth. Parnell. Babington. Hooker. Lindley.

PLATE XV.—B.

Calamagrostis epigejos,   Withering.
Arundo calamagrostis,   Smith. Knapp. Schrader.
"   "   Linnæus. Oeder. Ehrhart.
" *epigejos,*   Hudson.

The Purple-Flowered Small Reed.

Calamagrostis—Palm Grass.   *Lanceolata*—Lanceolate.

A BEAUTIFUL species, growing in fenny countries in moist shady situations.

A somewhat common Grass in England. Found also in Ireland, France, Germany, Switzerland, Italy, Spain, Portugal, Lapland, Norway, Sweden, Turkey, Greece, Siberia, British America, and North Africa.

Of no agricultural value.

Stem circular, smooth, striated, carrying four or five narrow, flat, acute, long, rough leaves, with smooth striated sheaths, the upper one extending beyond its leaf. Ligule of upper leaf bold and obtuse. Joints wide apart. Inflorescence compound panicled, seven to eight inches long, spreading when in flower, branches into clusters, rough, and slender. Spikelets numerous, composed of two narrow acute equal-sized glumes, and one floret shorter than the glumes; of two paleæ, the exterior being five-ribbed; apex bifid, awned from slightly below the apex. Base of palea hirsute, extending beyond the floret. Glumes destitute of lateral ribs, dentate for the whole length of their keel, and purple in colour. Awn rough, slender, very short. Inner palea one third shorter than the outer one, thin, transparent, and apex cloven. Styles two. Filaments three. Stigmas long and feathery. Anthers long, and cloven at either extremity. Scales acute. Length three feet. Root perennial and creeping.

Flowers in June and July, and ripens seed at the end of August. Distinguished from *C. epigejos* in the awn being very short.

The *C. Lapponica* of Ireland is looked upon as a variety of this species, and not the true *C. Lapponica* of Lapland.

The illustration is from a specimen gathered by Mr. J. Sidebotham, at Rosthern Mere, Cheshire.

CALAMAGROSTIS STRICTA.

NUTTALL. PARNELL. KOCH. LINDLEY. HOOKER.

PLATE XVI.—A.

Arundo stricta, SMITH. HOOKER. SCHRADER.
" *neglecta,* EHRHART.

The Small Close Reed.

Calamagrostis—Palm Grass.   *Stricta*—Upright.

THIS very rare Grass grows on bogs and in marshes. Found about Oakmere, in Delamere Forest, Cheshire, and in several places in the county of Antrim. It used to grow in Fifeshire, but has been by drainage of the land destroyed in that county. Found in the most northern portion of Europe, and in North America.

Stem upright, circular, and rough, having two or three narrow acute rough leaves, with smooth striated sheaths and smooth joints. Inflorescence compound panicled. Panicle somewhat close and rough, three to five inches in length. Calyx having two almost equal-sized membranous broad glumes, destitute of lateral ribs. Floret one-awned, consisting of two paleæ, ovate in form, apex jagged, and base hirsute; the awn does not extend beyond the apex of the palea. Inner palea considerably shorter, thin, and transparent. Length from eighteen to twenty-four inches.

It is much less robust than *Calamagrostis epigejos*, having shorter hairs, and the floret only half the size. Awn commencing lower, and not extending much beyond the palea.

Comes into flower about the 20th. of June, and ripens its seed in the last week in July.

AGROSTIS CANINA.

Linnæus. Smith. Hooker. Greville. Koch. Parnell.
Willdenow. Leers. Hoffmann.

PLATE XVI.—B.

Agrostis vinealis,   Withering.
"   *stricta*,   Sinclair.
Trichodium caninum,   Lindley. Schrader.
Agrostis tenuifolia,   Curtis.
"   *fascicularis*,   Sinclair.

The Brown Bent Grass.

Agrostis—A field (from the Greek).   *Canina*—Dogs.

Agrostis. *Linnæus.*—*Agrostis*, or Bent Grass, is derived from the Greek of *a field*, in reference to the species (of which there are six in Great Britain) growing in open situations. The panicle is loose, with laterally-compressed spikelets. Two acute awnless membranaceous glumes. Sessile floret. Two unequal-sized glumellas.

A COMMON species, found in boggy situations in England, Scotland, Ireland, France, Italy, Germany, Denmark, Sweden, and America. Of no agricultural value.

Stem circular, polished, slender, erect, with the base somewhat decumbent, carrying four or five taper-pointed narrow leaves with smooth sheaths, the upper one extending considerably beyond its leaf, and having at its apex a lengthy-pointed membranous ligule. Joints smooth. Inflorescence yellowish brown, compound panicled. Panicle upright, spreading when in flower. Branches slender, elastic, rough, mostly in clusters of three or five. Spikelets small, acute, numerous, on foot-stalks. Calyx of two glumes, the exterior one being destitute of lateral ribs, dentate the entire length of its keel, and larger than the inner glume. Floret consisting of one palea, ovate, five-ribbed, hairy at the base, dentate at the summit, and having a lengthy dorsal awn commencing from a little above the base, and extending half its length beyond the apex of the palea. Length from one to two feet. Root perennial and creeping.

There is a small alpine variety found on the Scotch mountains, which is only from two to three inches in length.

I am indebted to Mr. Sidebotham, of Manchester, and Dr. Wilson, of Nottingham, for specimens of this Grass.

AGROSTIS SETACEA.   A. ALBA.

AGROSTIS SETACEA.

Curtis. Hooker and Arnott. Smith. Parnell. Knapp. Babington. Lindley. Withering.

PLATE XVII.—A.

Agrostis alpina,	Withering. Hull.
" *canina,* var.	Hudson.
" *mutabilis,*	Sibthorp.

The Bristle-leaved Bent Grass.

Agrostis—A field.   *Setacea*—Bristle-like.

AN interesting very local species, confined to the dry downs of the south-west of England; being most abundant in Hampshire, Devonshire, and Cornwall, on sandy heaths, where it flourishes and finds food for flocks of sheep.

It is a native of France, Italy, Switzerland, Germany, Spain, Portugal, Turkey, and Greece.

Root perennial, tufted, and fibrous. Stem circular, rough, and striated; carrying four or five very narrow rough leaves, with striated sheaths, the uppermost considerably longer than its leaf. Joints three. Leaves from the root long, setaceous, and crowded. Inflorescence compound panicled, upright and compact until in flower, then spreading. Spikelets small, acute, and numerous, consisting of two almost equal-sized glumes, and one floret shorter than the glumes. The floret consisting of two unequal-sized paleæ, the exterior one four-ribbed, base hirsute, summit jagged, having an awn as long again as the palea, arising from slightly above the base, and being rough and slender. Inner palea diminutive. Styles two. Stigmas feathery. Filaments three. Anthers cloven at either extremity. Length from eight to fifteen inches.

Flowers in July, and ripens its seed at the beginning of September.

It is readily distinguished from other species. *Agrostis vulgaris* has stem and sheaths smooth, and inner palea half the length of outer one. *A. alba* has mostly no awn, and the leaves from the root are not setaceous. *A. canina* has smooth leaves and sheaths, and no inner palea.

The illustration is from a specimen gathered at Penzance, by Mr. J. Ralfs.

AGROSTIS ALBA.

Linnæus. Hooker and Arnott. Parnell. Smith.
Withering. Hull. Relhan. Sibthorp. Abbot. Willdenow.
Cullum. Schrader. Greville. Lindley.

PLATE XVII.—B.

Agrostis stolonifera,	Linnæus. J. E. Smith. Martens.
" "	Willdenow. Knapp. Koch.
" *compressa,*	Willdenow.
" *sylvatica,*	Linnæus. Hudson.
" *polymorpha,*	Hudson.
" *mutabilis,*	Knapp.
" *palustris,*	Sinclair.
" *capillaris,*	Leers.
" *stolonifera latifolia,*	Sinclair.

The Marsh Bent Grass.

Agrostis—A field.   *Alba*—White.

ABUNDANT in pastures and on wood sides, preferring a dry sandy soil, sometimes found two thousand feet above sea level. Of no use to the agriculturist.

Root perennial, tufted, and creeping. Stem circular, polished, and upright, bearing four or five short, narrow, flat, very rough leaves, with somewhat rough, striated sheaths, the upper one extending beyond its leaf, having at its apex a long, ragged, acute, ribbed ligule. Joints smooth. Inflorescence compound panicled. Panicle upright, somewhat purple, with pale green florets. Branches rough, slender, and spreading when in flower, proceeding from the rachis, generally in fives, placed at equal distances, but unequal in length. Spikelets small, upright, numerous, consisting of one small awnless floret. Calyx consisting of two nearly equal-sized, narrow, acute glumes, destitute of lateral ribs.

Floret of two unequal-sized paleæ, exterior one ovate, hairy at the base, and notched at the apex; inner one only half the length, with cloven apex, entire margins, and semitransparent.

Length from eighteen to twenty-four inches.

Flowers in the third week of July, and seeds ripen at the end of August.

This species is subject to variety.

Dr. Parnell describes two varieties, namely:—

1st.—*Stolonifera*. Growing on damp heavy ground near the sea, and by the side of ditches and other wet situations. It has the branches of the panicles densely tufted.

2nd.—*Palustris*. With larger spikelets, growing in damp, shady, stagnant situations.

Sir W. Hooker remarks, "In some there is a short awn at the base of the outer glumella; this constitutes the *Agrostis compressa* of Willdenow, and occasionally the flowers are viviparous, when it is the *A. sylvatica* of Linnæus."

The illustration is from a specimen forwarded by Mr. J. Sidebotham, of Manchester.

AGROSTIS VULGARIS.   A. SPICA-VENTI.

AGROSTIS VULGARIS.

WITHERING. HOOKER AND ARNOTT. PARNELL. SMITH.
HULL. RELHAN. KNAPP. SINCLAIR. SCHRADER. HOFFMANN. LINDLEY.
GREVILLE. KOCH.

PLATE XVIII.—A.

Agrostis canina,	WITHERING.
" *pumila*,	LIGHTFOOT. LINNÆUS.
" "	WILLDENOW. EHRHART.
" *tenuis*,	SIBTHORP.
" *capillaris*,	ABBOT. ROTH.
" *polymorpha*,	HUDSON.
" *hispida*,	WILLDENOW.
" *stolonifera*,	LEERS. EHRHART.

The Fine Bent Grass.

Agrostis—A Field.   *Vulgaris*—Common.

A COMMON English Grass, growing in dry pastures and on heaths. Of no agricultural value.

Common in Ireland and Scotland, and is a native of France, Italy, Germany, Norway, Sweden, Denmark, Lapland, and North America.

Root perennial, tufted, and creeping. Stem upright, circular, and polished, having five or six short, flat, rough, narrow, acute, dentato leaves, with smooth striated sheaths, the uppermost one slightly longer than its leaf, having an abrupt, blunt, membranous ligule at the apex. Joints smooth. Inflorescence compound panicled. Panicle upright, either brownish purple or light green. Branches spreading zigzag, delicate, rough, proceeding from the rachis at equal distances in threes or fours. Spikelets small, shining, numerous; of one awnless floret. Calyx of two almost equal-sized narrow acute glumes, destitute of lateral ribs. Floret of two unequal-sized paleæ; exterior one ovate, with smooth base and notched apex; inner one membranous; margins entire, and only half the length of the outer palea. Length fifteen inches.

Flowers at the commencement of July, and ripens seed in the middle of August.

Subject to variety.

The variety *pumila* is a pretty dwarf plant, growing in dry alpine situations. I found it abundant in Westmoreland and Cumberland—on Loughrigg Fell and Helvellyn, growing by the side of *Nardus stricta*. It does not exceed three inches in height.

The variety *aristata* has a long, slender, dorsal awn, arising a little above the base of the outer palea.

The illustration of this beautiful Grass is from a specimen sent by Mr. Joseph Sidebotham, of Manchester.

AGROSTIS SPICA-VENTI.

Linnæus. Hooker and Arnott. Smith. Hudson. Leers. Koch.
Oeder. Withering. Hull.
Relhan. Abbot. Willdenow. Knapp. Schrader.

PLATE XVIII.—B.

Anemagrostis spica-venti,   Parnell. Lindley.
Gramen harundinaceum,   Gerarde.

The Spreading Silky Bent Grass.

Agrostis—A Field.   *Spica-venti*—Wavy spike.

THIS beautiful Grass is by no means a common species, although it has been procured in the counties of Kent, Surrey, Middlesex, Hertfordshire, Essex, Suffolk, Norfolk, Cambridgeshire, Bedfordshire, Berkshire, Warwickshire, Yorkshire, Lancashire, Cumberland, Durham, and Northumberland. In Scotland it is one of the rarest Grasses, being only found on the Fifeshire coast.

Abroad it is procured in the middle and south of Europe.

Grows in light sandy soil, more particularly in fields that are sometimes flooded.

Root annual and fibrous. Stem upright, smooth, circular, carrying five narrow, acute, spreading, rough, ribbed leaves, with roughish sheaths; the upper one extending beyond its leaf, and having a lengthy, lanceolate, jagged ligule at its apex. Joints naked. Inflorescence compound panicled, spreading, and loose. Panicle of great size, glossy, with slender, rough, sub-divided branches disposed in alternate clusters, the centre one being the longest. Rachis usually smooth. Spikelets numerous, diminutive, of one awned floret of the same length as the calyx. Calyx consisting of two unequal acute glumes, with rough keels, the uppermost one largest. Floret of two paleæ, exterior one ovate-lanceolate, roughish, and bearing a tuft of hairs at the base.

Awn long, rough, slender; proceeding from slightly below the summit of the palea, and being above three times its length. Inner palea shorter. Apex bifid. Margins entire, membranous, and linear. Seeds exceedingly smooth.

Flowers in June and July.

This Grass is readily recognised by the great length of the awn, in comparison with that of the floret.

For specimens I am indebted to Mr. Joseph Sidebotham, of Manchester.

The illustration is from a specimen procured at Godalming, in the county of Surrey, by Mr. J. D. Salmon.

CATABROSA AQUATICA.

Beauvois. Hooker and Arnott. Parnell. Lindley.

PLATE XIX.

Aira aquatica,	Linnæus. J. E. Smith. Hooker.
" "	Withering. Hull. Sibthorp.
" "	Relhan. Hooker. Abbot.
" "	Curtis. Knapp. Ehrhart.
" "	Willdenow. Schrader. Host.
" "	Oeder. Greville.
Poa dulcis,	Salisbury.

The Water Whorl-Grass.

Catabrosa—A Gnawing.   *Aquatica*—Water.

Catabrosa. *Beauvois.*—*Catabrosa* or Whorl-Grass, has a spreading panicle with awnless florets. The name is derived from the Greek, and signifies *a gnawing*, on account of the extremity of the glumes being erose.

Catabrosa aquatica is the only British species.

A BOLD handsome species, growing in ditches, banks of rivers, and other wet situations, rendering it a useless agricultural Grass, although cattle are very fond of it.

A common Grass in England, Scotland, and Ireland. Abroad it is a native of France, Italy, Germany, Sweden, Norway, Lapland, and in the north of South America.

Root perennial, extremely long, branched, creeping, or frequently floating on the water, and having lengthy shining white fibres. Stem strong, circular, and smooth, the basal portion procumbent and floating in the water, the upper portion rising from twelve to eighteen inches above the water, bearing three or four broadly-linear leaves, having striated, smooth, lax sheaths, the upper one shorter than its leaf, and having a blunt membranous ligule at its apex. Joints smooth. Inflorescence compound panicled, the panicle being upright, with spreading unequal branches. Spikelets small, numerous, and weeping, consisting of two awnless florets, considerably longer than the glumes. The calyx consisting of two rough, unequal blunt, membranous glumes, destitute of lateral ribs. The upper glume largest. Florets of two equal-sized paleæ. Length from twelve to twenty-four inches.

Flowers the second week of July, and ripens seed the second week of August.

A dwarf variety has been found growing near the sea in damp situations, especially along the west coast of Scotland. This variety is known by the English name of "Small Water Hair Grass," and is the *Catabrosa aquatica*, variety *littoralis*, of Parnell.

I am indebted to Mr. J. Sidebotham for specimens.

The specimen illustrated was gathered in a ditch, near the railway station, Beeston, Nottinghamshire.

AIRA CÆSPITOSA.
XX

AIRA CÆSPITOSA.

LINNÆUS. PARNELL. SMITH. HOOKER. ARNOTT. GREVILLE. KNAPP.
WILLDENOW. SCHRADER. LEERS. HOST. EHRHART.
OEDER. HUDSON. WITHERING. HULL. SIBTHORP. ABBOT. RELHAN.

PLATE XX.

Deschampsia cæspitosa,   LINDLEY.
Gramen segetale,   GERARDE.

The Tufted Hair-Grass.

Aira—To destroy.   *Cæspitosa*—Tufted.

AIRA. *Linnæus.*—The Hair-Grass is named from the Greek, and signifies to destroy, but why it received this unwelcome name is apparently uncertain. There are six British species. Having a spreading panicle, of which the spikelets are laterally compressed. There are two florets present in each spikelet, with a third imperfect rudiment between them; the outer palea of each floret is rounded at the back and furnished with an awn.

A VERY handsome Grass, the flowers of which are well adapted for decoration, being very graceful. It is a common species in England, Scotland and Ireland, of no agricultural merit, being coarse and rough, and with but little nutritive properties. It will flourish in almost any situation, but prefers damp fields, where it grows into large tufts, and is known to agriculturalists as hassocks, a Grass difficult to destroy.

It is a native of Norway, Sweden, Lapland, France, Italy, Germany, North America, and the United States. Nowhere does it flourish so luxuriantly as on the banks of a brook.

The plant forms a large coarse tuft, and, as it is not eaten by cattle except when nothing else can be procured, a field in which it abounds has a singularly unsightly, and to farmers unwelcome appearance.

The root is perennial and fibrous. Stem upright, circular, and rough, and supporting four or five narrow, rough, coarsely-ribbed, acute leaves with rough striated sheaths, the uppermost one extending considerably beyond its leaf, and having a lengthy, membranous, acute ligule at its apex. Joints smooth and very strong.

Inflorescence compound panicled, and exceedingly handsome. Panicle when first expanded drooping, afterwards becoming upright, with the branches spreading in all directions. Branches and rachis rough. The spikelets are small and numerous, consisting of two or three horned florets. The calyx consisting of two equal-sized acute glumes, the upper one three-ribbed and the lower one destitute of lateral ribs. Of the two paleæ the exterior one of the lowest floret not equal in length to the glumes, membranous, base hirsute, destitute of lateral ribs, jagged on the summit; a slender awn starts from a little above the base as far as the apex of the palea. The inner palea rather shorter, linear, margin entire, and also membranous. Length from two to five feet.

Aira alpina differs, on account of the awn commencing in that species slightly above the centre of the outer palea; whilst on the other hand *Aira flexuosa* has the awn of the lower floret protruding above one third its length beyond the glumes.

There is a viviparous variety of *Aira cæspitosa*, known under the name of variety *vivipara*, which has been found on the Clova Mountains. It is an interesting ornamental variety, worthy of extensive cultivation, as a border flower in gardens. Dr. Parnell describes two varieties, namely, "*longiaristata*" and "*brevifolia.*" The variety *longiaristata* differs in having the awn of the outer palea extending one quarter of its length beyond the apex of the palea, and in having the spikelets of a rich chocolate colour. Found on the mountains in Perthshire between three and four thousand feet above the sea level. The variety *brevifolia* is distinguished by the very short radical leaves. It is found growing near the summit of several of our highest Scotch mountains.

This species usually flowers in the beginning of July, and ripens its seed early in September.

Door-mats and basses are made of the hay of *Aira cæspitosa*, and for this the Grass is much in repute by rural cottagers.

The illustration is from a plant growing on the edge of the lake at Highfield House.

AIRA ALPINA.   A. CARYOPHYLLEA.

AIRA ALPINA.

Linnæus. Hooker and Arnott. Parnell. Babington. Lindley. Willdenow. Wahlenberg.

PLATE XXI.—A.

Aria lævigata,   J. E. Smith.

The Smooth Alpine Hair-Grass.

Aira—To destroy.   *Alpine*—Mountain.

AN uncommon and useless Grass, not seen at a less elevation than three thousand feet.

Found on Ben Lomond, Ben Arthur, and moist rocks in Angusshire, and said to be found in Wales.

Native of Lapland, Scotland, and North America.

Distinguished from *Aira flexuosa* by the awn rising from above the centre of the palea, and not extending beyond the apex of the palea.

Stem upright, circular, and polished, carrying three or four narrow, acute, mostly involute, strongly-ribbed leaves, rough on inner surface and margins, smooth on back, with smooth striated sheaths. Joints smooth. Inflorescence compound panicled. Panicle upright, silky, brown. Apex drooping. Branches arranged on the smooth rachis in pairs at certain distances. Spikelets numerous, with exceedingly delicate footstalks, usually two, though occasionally three, awned florets, the lower one not protruding beyond the calyx. Calyx of two nearly equal membranous smooth glumes. Upper glume three-ribbed, others destitute of lateral ribs. Florets of two paleæ, exterior one of lowest floret shorter than the glumes; oval in shape; base hirsute; apex jagged. Keel somewhat rough, having a brief rough awn rising from a little above the centre, and extending to the apex of the palea. Inner one rather shorter and membranous. Root perennial and fibrous. Length from twelve to eighteen inches.

Flowers at the commencement of August, and becomes ripe in the middle of September.

Dr. Parnell figures a viviparous variety called *vivipara*.

AIRA CARYOPHYLLEA.

Linnæus. Hooker and Arnott. J. E. Smith. Parnell. Babington.
Reichenbach. Greville. Lindley.
Willdenow. Knapp. Curtis. Stillingfleet. Greaves. Oeder.

PLATE XXI.—B.

Avena caryophyllea,   Koch.

The Silver Hair-Grass.

Aira—To destroy.   *Caryophyllea*—........?

ANOTHER useless agricultural grass, growing in dry gravelly situations, and tolerably abundant in England, Scotland, and Ireland.

It is a very handsome species.

Found in Germany, France, and Italy.

Stem upright, circular, smooth, and striated, bearing three or four short, narrow, rough leaves, with striated rough sheaths, upper leaf much longer than its sheath, and having a prominent acute ligule at the apex. Joints smooth. Inflorescence compound panicled, silvery grey. Panicle upright, triple-forked, spreading, tinged with purple. Rachis smooth. Spikelets small, with rounded bases and slightly swollen, consisting of two awned florets, not protruding beyond the apex of the glumes. Calyx of two equal-sized membranous glumes, destitute of lateral ribs. Florets of two equal-sized paleæ, exterior one of lowest floret bifid, base hairy, furnished with a slender awn rising from slightly beneath the centre, and extending half its length beyond the apex of the palea. Inner one thin and membranous. Root annual and fibrous. Length from six to twelve inches.

Flowers in the third week of June, and ripens its seed at the end of July.

This Grass is common in Sherwood Forest, where the specimen from which the illustration is taken was gathered.

AIRA FLEXUOSA.

AIRA FLEXUOSA.

Linnæus. Smith. Hooker and Arnott. Parnell. Babington.
Reichenbach. Koch. Greville. Willdenow.
Knapp. Schrader. Leers. Host. Schreber. Oeder. Ehrhart.

PLATE XXII.

Aira montana,	Hudson. Dickson.
" "	Leers, (*not* Linnæus.)
" *scabro-setacea,*	Knapp.
" *setacea,*	Hudson.

The Wavy Mountain Hair-Grass.

Aira—To destroy.   *Flexuosa*—Bending.

A COMMON Grass in England, Scotland, and Ireland, on heaths and upon hills, often growing amongst the heather.

It is eaten by sheep.

Abroad it is found in North America, France, Italy, Norway, Sweden, Lapland, and Germany.

The spikelets are twice as large as in *Aira caryophyllea.*

Stem upright, smooth, striated, rather flat, bearing three or four exceedingly narrow, long, smooth leaves, and numerous radical ones, with roughish striated sheaths, the uppermost one considerably longer than its leaf, and crowned with an acute membranous ligule. Joints smooth. Inflorescence compound panicled; colour pale brownish green. Panicle upright, with delicate, rough, triple-forked, spreading branches. Spikelets upright, of two awned florets, which do not protrude beyond the calyx; colour brownish glossy copper. Calyx consisting of two almost equal-sized membranous glumes, with somewhat rough keels, but destitute of lateral ribs. Florets of two equal-sized paleæ, the exterior one of lowest floret having a bifid apex; base hirsute, with two delicate ribs on either side. Keel roughish, and having a slender awn rising from slightly above the base, and extending far beyond the apex of the palea. Inner palea very thin, membranous; margins minutely fringed. Root perennial and fibrous. Length from twelve to eighteen inches. Colour dark green.

Flowers at the commencement of July, and ripens its seed in the middle of August.

Dr. Parnell describes a variety known as *A. flexuosa*, var. *montana*. Frequently met with on the Highland Moors. It is more slender, and the ligules are more acute.

The illustration is from a specimen forwarded by Mr. Joseph Sidebotham, of Manchester.

AIRA CANESCENS.

Linnæus. Hooker and Arnott. J. E. Smith. Parnell. Schrader.
Knapp. Willdenow. Dickson. Withering. Ehrhart. Oeder.

PLATE XXIII.—A.

Corynephorus canescens, Beauvois. Babington.
" " Reichenbach. Koch. Kunth.
Gramen junceum, Dalechamps.

The Grey Hair-Grass.

Aira—To destroy.   *Canescens*—To become grey.

ONE of the rarest of the British Grasses, and consequently a useless agricultural species.

Found on the sandy coasts of Norfolk, Suffolk, Dorset, and Jersey. Native of the Islands of the Mediterranean, Greece, Turkey, Spain, Portugal, Italy, Switzerland, Holland, Germany, Belgium, France, Italy, Norway, and Sweden.

Easily distinguished from all other British species in having club-shaped awns, which are fringed in the centre.

Stem upright, circular, and smooth, bearing four or five setaceous, very short, rough, and glaucous leaves, with rough striated sheaths, the uppermost leaf shorter than its sheath. Ligule of upper leaf acute and bold. Joints three, the uppermost one naked. Inflorescence compound panicled, close and compact until in flower, then spreading; having a purple tinge. Branches rough, but rachis smooth. Spikelets consisting of two acute, membranous, equal-sized glumes, destitute of lateral ribs; keels minutely dentate, and two florets shorter than the glumes. The florets composed of two equal-sized paleæ, the exterior one acute, base hairy, and without lateral ribs, and having a lengthy dorsal awn. Inner palea membranaceous and narrow; apex notched. Awn rising from a little above the base of the exterior paleæ, and extending half its own length beyond; club-shaped above, and having a circular fringe in the centre. Styles short, two. Stigmas long and feathery. Filaments slender, three in number. Anthers dark purple in colour, and short. Root annual or biannual, and fibrous. Length from six to fourteen inches.

Flowers in July, and ripens its seed in August.

AIRA PRÆCOX.

Linnæus. Hooker and Arnott. J. E. Smith. Parnell. Babington.
Lindley. Greville. Willdenow.
Curtis. Knapp. Graves. Schrader. Oeder. Ehrhart.

PLATE XXIII.—B.

The Early Hair-Grass.

Aira—To destroy.   Præcox—Early.

THIS is a very early Grass, and of but little value; it grows on sandy hills and wall tops.

Tolerably abundant in England and Ireland, less common in Scotland.

Found in France, Italy, Germany, and North America.

This species is known from *Aira caryophyllea* by the close panicle, which does not exceed half an inch in width.

Stem circular, smooth, upright, and carrying four or five narrow roughish leaves, with rough, striated, somewhat inflated sheaths. Uppermost sheath longer than its leaf, and having at its apex a lanceolate membranous ligule. Joints smooth. Inflorescence simple panicled; greenish silvery colour. Panicle upright and close; branches rough; rachis smooth. Spikelets of two awned florets, enclosed within the calyx. Calyx composed of two equal and acute glumes, minutely toothed on the keels, but destitute of lateral ribs. Florets of two equal-sized paleæ; exterior one of lowest floret bifid, base hairy, obscurely five-ribbed, and having a lengthy, rough, slender awn rising from slightly above the base, and extending half its length above the apex of the palea. Inner one with margin minutely fringed, and membranous. Root annual and fibrous. Length from four to six inches.

Flowers at the end of May, and is ripe in a month.

The illustration is from a specimen gathered in Dunham Park, by Mr. Joseph Sidebotham, of Manchester.

MOLINIA CÆRULEA.
XXIV

MOLINIA CÆRULEA.

MŒNCH. HOOKER AND ARNOTT. PARNELL. LINDLEY. KOCH. STURM.
BEAUVOIS. BABINGTON. REICHENBACH.

PLATE XXIV.

Melica alpina,	DON.
Molinia depauperata,	LINDLEY. PARNELL.
Melica cærulea,	LINNÆUS. SMITH. HOOKER. HOST.
" "	GREVILLE. WILLDENOW. SCHRADER.
" "	CURTIS. KNAPP.
Aira cærulea,	LINNÆUS. HUDSON. LEERS. OEDER.

The Purple Molinia.

Molinia—After Molina, a Naturalist.   *Cærulea*—Blue.

MOLINIA. *Mœnch.*—A genus named after Don Giovanni Ignatio Molina, who published a work upon the Natural History of Chili seventy-five years ago. There is only one British example, namely, *Molinia cærulea.*

THIS reed-looking Grass, perhaps better known as the Purple Melic Grass, is abundant throughout Scotland, England, and Ireland, growing on damp heathy moors. It is of but little agricultural value. In the Orkney and Shetland Islands the stems are made by the fishermen into ropes, whilst in England they are manufactured into cheap brooms.

It occurs in Lapland, Norway, and Sweden, and to the most southern portions of Europe.

Stem upright, circular, smooth, being bulbous at the base, carrying three lengthy, narrow, linear, taper-pointed, acute, rough leaves, with smooth striated sheaths, the uppermost one shorter than its leaf, bearing at its apex a diminutive ligule. Joints smooth, and close to the base. Inflorescence compound panicled. Panicle upright, lengthy, narrow, and compact; branches slender, rough, and situated on the rachis in bunches at certain intervals. Spikelets numerous, small, chiefly composed of two, yet sometimes of three, awnless florets, purplish in colour, and considerably longer than the glumes. Calyx smooth, of two unequal, acute glumes. Florets of two equal-sized paleæ, exterior one of basal floret smooth, three-ribbed, and acute; inner one having two bold marginal ribs of a green colour. Root consisting of a multitude of strong fibres, perennial. Length from one to two feet and a half.

Flowers towards the end of July, and ripens its seed at the close of August.

There is a variety known as *Molinia depauperata*, which Lindley and Parnell gives as a distinct species, and known as the Tawny Melic Grass. It differs in being usually less in size, and having the leaves of the stem extending beyond the panicle, in the calyx having only one floret, the outer palea being five-ribbed. It was discovered on the Clova Mountains, (three thousand feet above the sea,) by Mr. Donald Munro. There is another variety more dwarf and compact, known as *M. cærulea*, var. *breviramosa*, distinguished by its dwarf habit and dark purple inflorescence—a common Grass on moors.

The specimen for illustration was forwarded by Mr. Joseph Sidebotham, of Manchester.

MELICA NUTANS.

LINNÆUS. HOOKER AND ARNOTT. SMITH. PARNELL. GREVILLE.
LINDLEY. KOCH. WILLDENOW. CURTIS. MARTYN. KNAPP.
RELHAN. GRAVES. SCHRADER. BABINGTON. HOST.
SCHREBER. LEERS. REICHENBACH. WITHERING. HULL. DICKSON.

PLATE XXV.—A.

Melica montana,   HUDSON.
Poa nutans,   HALLER.

The Mountain Melic Grass.

Melica—Honey.   *Nutans*—Nodding.

MELICA, *Linnæus.*—An interesting family, of which there are but two British examples. The name is derived from *mel*—honey.

A GRASS as yet of no agricultural value, growing in damp shady woods at an altitude of about five hundred feet above the sea, and not found higher than two thousand feet. It is an early Grass, doing well under cultivation, and therefore may prove of use to the farmer.

In Scotland it occurs in Aberdeenshire, Forfarshire, Fifeshire, and near Edinburgh. In England, in all the northern counties and Nottinghamshire, Derbyshire, Worcestershire, Suffolk, and Hertfordshire. In Wales, in Denbigh. Abroad, it is a native of France, Italy, Germany, Denmark, Sweden, Norway, and Lapland.

This lovely Grass is very ornamental when growing luxuriantly, and no one can fail being struck with its beauty after seeing it growing, as it does, in a damp wood near Ambleside. The wood seemed as if meant for Fairyland, each raceme of bloom bearing a number of bells, all hanging in one direction.

Stem upright, slender, roughish, bearing four or five long, narrow,

M

acute, flaccid, pale green leaves, with rough striated sheaths, the upper one shorter than its leaf, and having at the apex a brief obtuse ligule. Inflorescence racemed. Raceme long, mostly of ten spikelets, placed on short rough footstalks. Spikelets large, ovate, pendulous, and consisting of two perfect and one imperfect floret. Calyx of two broad, reddish brown, smooth, five-ribbed glumes. Florets of two paleæ. Length from twelve to twenty-four inches. Some specimens gathered near Ambleside were above three feet in length. Root perennial and creeping.

Flowers at the end of May, and becomes ripe in July.

This Grass is known from *Melica uniflora* in the inflorescence being racemed instead of simple panicled, and in the calyx containing two perfect florets. The most unaccustomed eye can at once recognise the difference between these two species.

My thanks are due to Mr. Joseph Sidebotham for specimens gathered near Halifax, and from which the illustration is taken.

MELICA UNIFLORA.

LINNÆUS. HOOKER AND ARNOTT. SMITH. PARNELL. GREVILLE.
WITHERING. LINDLEY. KOCH. HULL. RELHAN. ABBOT.
SIBTHORP. CURTIS. DICKSON. MARTYN. REICHENBACH. BABINGTON.
RETZIUS. WILLDENOW. KNAPP. GRAVES. SCHRADER. OEDER.

PLATE XXV.—B.

Melica nutans,   HUDSON. RUDBECK.
" *Lobelii,*   VILLARS.

The Wood Melic Grass.

Melica—Honey.   *Uniflora*—One-flowered.

A GRASS of but little agricultural value, flourishing in clayey soil in damp rocky woods.

A frequent Grass in England, Ireland, Scotland, Italy, France, and Germany.

This beautiful species is very ornamental, and when dried is well adapted for winter decoration.

Stem upright, circular, and slender, bearing four or five long, flat, thin, acute, flaccid, rough leaves, with rough striated sheaths, whose upper portions are furnished sparingly with slender yet conspicuous white hairs. Upper sheath shorter than its leaf, and having at its apex a short membranous ligule. Inflorescence simple-panicled. Panicle slightly pendulous, with few spikelets on long, slender, rough footstalks, the branches long and slender, rising usually in pairs from the rachis. Spikelets upright, oval in shape, consisting of a perfect and an imperfect awnless floret hid within the calyx. Calyx of two reddish brown five-ribbed, smooth glumes. Florets of two paleæ. Length from twelve to eighteen inches. Root perennial and creeping.

Flowers in the middle of June, and becomes ripe at the end of July.

It is distinguished from *Melica nutans* in the simple panicle, and in the calyx containing only one perfect floret.

This species flourishes luxuriantly about Ambleside, and also in a wood near Critch, in Derbyshire; where it grows well it is very beautiful.

The illustration is from a Derbyshire specimen.

HOLCUS MOLLIS.

Linnæus. Hooker and Arnott. Smith. Parnell. Lindley.
Koch. Willdenow. Curtis. Knapp. Sinclair.
Schrader. Leers. Host. Schreber. Babington. Hudson. Abbot.
Withering. Relhan. Hull. Sibthorp. Dickson.

PLATE XXVI.

The Creeping Soft Grass.

Holcus—To extract.   *Mollis*—Soft.

Holcus. *Linnæus.*—Of the genus *Holcus*, or "Soft Grass," England can boast of only two species, both exceedingly interesting plants. The name is derived from the Greek, and signifies to *extract*; the genus taking this singular name because it was supposed to have the property of drawing out thorns from the flesh.

A GRASS of no agricultural value, as cattle refuse to eat it, whilst its long creeping roots speedily impoverish the soil. Its favourite habitat is sandy, light, barren soil.

Common in Britain, Sweden, Denmark, France, Germany, and Italy.

Stem upright, circular, and smooth, bearing four or five flat, broad, acute, soft, roughish, pale green leaves, with usually smooth sheaths, upper sheath considerably longer than its leaf. Joints four, hairy. Inflorescence compound-panicled. Panicle upright, and slightly pendulous at the apex. Spikelets consisting of two florets, the upper one awned. Calyx consisting of two equal-length glumes, membranous, and keels hairy. Upper glume three-ribbed; lower one destitute of lateral ribs. Florets of two paleæ. From near the apex of the upper floret arises a long awn, which is rough from the base to the apex. Length from one to three feet. Root perennial and creeping.

Dr. Parnell describes two varieties, one *biaristatus*, which has larger and fewer spikelets; the other *parviflorus*, does not exceed twelve inches in length, and having very small spikelets.

The specimen for illustration was sent by Dr. Wilson, of Nottingham.

HOLCUS LANATUS.

HOLCUS LANATUS.

LINNÆUS. SMITH. PARNELL. HOOKER AND ARNOTT. LINDLEY.
KOCH. HUDSON. WITHERING. HULL. RELHAN. ABBOT.
SIBTHORP. CURTIS. DICKSON. REICHENBACH. BABINGTON. WILLDENOW.
KNAPP. SINCLAIR. LEERS. HOST. SCHREBER.

PLATE XXVII.

The Meadow Soft Grass.

Holcus—To extract.   *Lanatus*—Woolly.

A PRODUCTIVE Grass, easily cultivated, yet cattle do not like it. It seems to delight to grow in shady situations, especially in light moist soils.

Common throughout Great Britain and Scotland, France, Italy, and Germany.

This very beautiful Grass has an upright circular stem, bearing four or five pale green, flat, broad, acute, soft, hairy leaves, with soft downy sheaths, the upper sheath extending considerably beyond its leaf; inflated and having at its apex an obtuse membranous sheath. Joints, four, hairy. Inflorescence compound-panicled, green, red, or pink in colour. Panicle upright, triangular in shape, compact and close when young, and spreading when more mature. Branches hairy. Spikelets pendulous. Two florets, the upper one awned. Calyx consisting of two hairy membranous glumes, the upper one oblong, tipped with a minute bristle. Keel hairy, having a green rib on either side; lower glumes crescent-shaped, and destitute of lateral ribs. Two equal-sized paleæ. Upper floret smallest, and elevated on a lengthy naked footstalk, having a dorsal awn of about half the length of the palea, commencing a little beneath the apex, and when mature curved in the form of a fish-hook. The apex of the awn is rough, but the lower two thirds is quite smooth. Length from twelve to twenty-four inches. Root perennial and fibrous.

This species is distinguished from *Holcus mollis* in having two thirds of the awn smooth.

Flowers at the beginning of July, and ripens its seeds in four weeks.

The specimen for illustration was forwarded by Dr. Wilson, of Nottingham.

ARRHENATHERUM AVENACEUM.

Beauvois. Hooker and Greville. Lindley. Babington.

PLATE XXVIII.

Arrhenatherum elatior,	Koch.
" *bulbosum,*	Lindley. Dunal.
Avena elatior.	Linnæus. Hudson. Curtis.
" "	Martyn. Cullum. Schreber.
" "	Leers. Withering. Relhan.
" "	Hull. Abbot.
" *nodosa,*	Cullum.
" *precatoria,*	Thuill.
Holcus avenaceus,	Smith. Hooker. Greville.
" "	Scopoli. Wiggers. Sibthorp.
" "	Knapp. Sinclair. Schrader.
" "	Reichenbach.
Gramen bulbosum nodosum,	Lobel.
" *caninum nodosum,*	Gerarde.

The Oat-like Grass.

Arrhenatherum—Male-awn.   *Avenaceum*—Of oats.

ARRHENATHERUM. *Beauvois.*—There is only a solitary example of this genus in this country, namely, the *Arrhenatherum avenaceum* It has the habit of the Oat Grass, yet differing in the number and structure of its florets. The name is derived from two Greek words, signifying *male* and *awn.*

A PRODUCTIVE agricultural Grass, especially on clayey soil, yet a species but little grown in this country. It will thrive well under trees, and is sometimes a troublesome weed in corn fields.

A common Grass in England, Scotland, Ireland, France, Germany, Italy, and the United States.

Stem upright, circular, and polished, bearing **four or five flat, rough, narrow, acute leaves**, with striated smoothish sheaths, **the** upper one **longer** than its leaf, and having at its apex a small ragged ligule. **Joints** smooth, and sometimes hirsute. Inflorescence simple panicled. **Panicle** inclining to one side, branches short and rough, the lower ones mostly in fives. Calyx consisting of a pair of very unequal acute membranous glumes. The florets consisting of two paleæ. The **lower** floret has a long awn rising from slightly above the base of the **outer palea,** the second floret has a very short awn commencing beneath **the apex.** Length **from two to three feet** and a half. Root perennial, **fibrous,** and sometimes **bulbous.**

There **is a variety known as** *bulbosum,* which grows in rich cultivated **fields, having bulbous** roots.

Flowers **in the** third week in **June,** and becomes ripe in about five **weeks.**

The present species, which is readily known from all other Grasses, **is a common** plant in Nottinghamshire, growing occasionally to the **height of** three feet and a half, especially in low meadows on the **banks of** hedges.

There **is** only another species **known of this family,** namely, the *A. pallens,* a Portuguese plant.

The specimen for illustration was forwarded by Mr. Joseph Sidebotham, **of** Manchester.

HIEROCHLOE BOREALIS.

ROEMER AND SCHULTES. PARNELL. HOOKER AND GREVILLE.
J. E. SMITH. LINDLEY. BABINGTON. REICHENBACH

PLATE XXIX.—A.

Holcus borealis,	SCHRADER.
" *odoratus,*	LINNÆUS. SMITH. WILLDENOW.
" "	OEDER. SINCLAIR. WAHLENBERG.
Hierochloe odorata,	KOCH.

The Holy Grass.

Hierochloe—Sacred Grass. *Borealis*—North.

HIEROCHLOE. *Gmelin.*—The "Holy Grass" has a wide-spread panicle, and derives its name from two Greek words, signifying *Sacred Grass*, because according to Gmelin, it is on the sacred festivals in some parts of Prussia, scattered before the doors of churches, being dedicated to the Virgin Mary. Sir W. Hooker remarks that a similar custom still prevails at Norwich, where the *Acorus calamus*, or "Sweet Sedge," is the favoured plant.

Great Britain only possesses one species, namely, *Hierochloe borealis*, which has been found in Scotland. It is an abundant Iceland plant.

THIS very rare species, although one of our earliest, yet it is not a valuable Grass.

It is confined to Scotland, having been found by the late Mr. G. Don, in a mountain valley called Kella, near the Spittle of Glen Shee, Forfarshire, and near Thurso, Caithness, in 1854, by Mr. Robert Dick.

Abroad it is native of Norway, Sweden, Lapland, Iceland, Italy, France, Germany, Prussia, Kamtschatka, and Russian America.

This is the Grass used for strewing before the Prussian churches. In Sweden it is hung over beds in the belief that it induces sleep.

In Iceland it is used to scent the clothes and apartments of the inhabitants, and in that island it is a common species. The scent emitted is very similar to that of our Sweet-Scented Vernal Grass, (*Anthoxanthum odoratum*.)

The stem stout, upright, circular, and smooth, carrying three or four wide, brief, lanceolate, roughish leaves, with smooth sheaths, the upper one somewhat swollen, considerably longer than its leaf, and having a bold broad ligule at its apex. Joints hid by the sheaths, near the base of the stem, and smooth. Inflorescence compound panicled. Panicle upright, except near the apex. Branches smooth, spreading, proceeding from the rachis in pairs; colour purplish. Spikelets of good size, glossy, green and purple, consisting of three awnless florets hid by the calyx. Calyx of two almost equal-sized, smooth, acute, broad glumes, destitute of lateral ribs. Florets of two paleæ; the exterior one of lowest floret five-ribbed; edges fringed; keel rough and slightly hairy; interior one shorter, with entire apex. Filaments in perfect floret two, in barren floret three. Anthers conspicuous, pendulous, and notched at either extremity. Ovarium ovate. Styles two. Stigmas feathery. Length from twelve to eighteen inches. Root creeping and perennial.

Flowers at the commencement of May, and ripens its seed in June.

The specimen for illustration was gathered near Thurso by Mr. R. Dick, and forwarded to me by Mr. Joseph Sidebotham, of Manchester.

KOELERIA CRISTATA.

Persoon. Hooker and Arnott. Koch. Babington.

PLATE XXIX.—B.

Aira cristata,	Linnæus. J. E. Smith. Hooker.
" "	Greville. Knapp. Schrader.
" "	Hudson. Reichenbach.
Airochloa cristata,	Link. Parnell. Lindley.
Poa cristata,	Willdenow. Withering. Hull.
" "	Relhan. Sibthorp. Abbot.
" "	Host. Leers. Ehrhart.

The Crested Hair-Grass.

Koeleria—After the continental botanist, Koeler.   *Cristata*—Crested.

KOELERIA. *Persoon.*—The present genus is named in honour of the author of a work on the Grasses of Germany and France, published fifty-five years ago by George Louis Koeler. The panicle is rounded and spiked.

England possesses only one species, namely, *Koeleria cristata*.

THIS Grass is rejected by cattle. It grows in dry situations near the sea, and on rocks as much as fifteen hundred feet above the sea-level. Frequent, more especially in the north of England, in Scotland, and in Ireland.

Native also of France, Italy, and Germany.

Better known as the *Aira cristata* of Linnæus.

Stem upright, circular, pubescent, bearing two or three stiff, rough, pubescent, narrow, acute, leaves, with hairy striated sheaths, the upper one longer than its leaf, having at its apex a short jagged ligule. Joints near the base smooth. Inflorescence simple-panicled, compact and silvery in appearance. Panicle upright, oval; sometimes two inches in length, interrupted near the base. Branches arranged in pairs on pairs on the rachis, close until in flower, then spreading. Spikelets compressed, consisting of two awnless florets, which do not extend beyond the glumes of the calyx. Calyx of two acute unequal-sized glumes, with dentate keels; uppermost glume three-ribbed. Florets of two equal-sized paleæ. Inner palea cloven at the apex; second floret having a lengthy downy footstalk. Length from three to six inches. Root perennial, forming dense tufts of lengthy downy fibres.

Flowers towards the end of June, and ripens its seed at the end of August.

The specimen for illustration was gathered at Castleton, by Mr. Joseph Sidebotham.

SESLERIA CÆRULEA.   PANICUM CRUS-GALLI.

SESLERIA CÆRULEA.

Scopoli. Hooker and Arnott. Parnell.
Koch. Smith. Reichenbach. Withering. Lindley. Knapp. Host.
Schrader. Oeder. Arduino. Hull.

PLATE XXX.—A.

Cynosurus cœruleus,	Linnæus. Willdenow. Hudson.
" "	Martyn. Ehrhart. Dickson.
" "	Jacquin. Wulfen.

The Blue Moor-Grass.

Sesleria—Named after the Italian botanist Sésler.   *Cærulea*—Blue.

Sesleria, *Linnæus.*—The genus *Sesleria* is named after another botanist, Leonard Sesler, an Italian Physician. Panicle spiked. There is only one British representative, the *Sesleria cærulea*, which is confined to the north.

A MOST beautiful Grass, yet of no agricultural merits. Found in the counties of Durham, Westmoreland, Cumberland, and Yorkshire, growing on the high hills. More abundant on the Scotch mountains, especially on Ben Lomond, (three thousand feet above the sea.) In Ireland it has been found, although rare, in the county of Sligo. Its favourite habitats are limestone rocks.

Native of France, Italy, Germany, Sweden, and Iceland.

Stem upright, circular, smooth, slender, pale green, carrying three very brief leaves, with smooth sheaths, the upper one eight times the length of its leaf, and having a short membranous ligule. Joints hid. Root leaves linear, narrow, lengthy, and eleven-ribbed. Margins minutely dentate. Inflorescence racemed. Raceme purplish blue, oval, and about half an inch in length. Spikelets laid one over the other all round the rachis. Calyx of two broad, acute, membranous, equal-sized glumes, destitute of lateral ribs. Florets of two equal-sized paleæ, exterior one of lowest floret five-ribbed. Apex dentate, and ending in a brief awn. Inner one linear bifid, having two green marginal ribs. Filaments three. Anthers conspicuous, notched at either extremity. Ovarium minute, hirsute, globose, with a pointed base. Style short. Stigmas conspicuous, long, linear, and pubescent. Length from six to twelve inches. Root creeping, having lengthy fibres, perennial.

Flowers at the end of April, and the seeds become ripe in the middle of June.

The specimen for illustration was gathered in Teesdale, by Mr. Joseph Sidebotham, of Manchester.

PANICUM CRUS-GALLI.

Linnæus. J. E. Smith. Hooker and Arnott. Lindley.
Koch. Willdenow. Knapp. Curtis. Graves. Schrader. Hudson.
Leers. Ehrhart. Withering. Hull.

PLATE XXX.—B.

Panicum vulgare,   Gerarde.
Echinochloa crus-galli,   Beauvois. Parnell. Babington.
"   "   Reichenbach.
Oplismenus crus-galli,   Kunth.

The Loose Panick Grass.

Panicum—Bread.   *Crus-galli*—........?

Panicum, *Linnæus.*—The Panick Grass, named after the Latin *Panis*—bread, from the circumstance that the seeds of some of the species are made into bread. Spikelets flat in front, and rounded on the back. There is only one British species, *Panicum crus-galli.*

A HANDSOME interesting species, although a strong coarse-growing plant, growing in damp situations, and of no agricultural use. There seems to be some doubt as to whether it is strictly British.

Found on waste land near Thetford, Norfolk; in fields near London. In Hampshire and Surrey.

A native of France, Germany, Italy, Switzerland, Belgium, Holland, Norway, Sweden, north of Africa, and the United States of America.

Stem upright, smooth, and striated, having three or four broad, pointed, ribbed (with marginal dentations) leaves, and smooth striated sheaths; upper sheath of same length as its leaf; no ligule. Joints three. Inflorescence compound-panicled, compact, secund; rachis angular; branches rough. Spikelets almost sessile, in clusters, composed of two glumes and two florets, one of the florets neutral. Glumes unequal, inner one considerably the largest, three-ribbed, and hirsute. Lowest floret barren. Styles two. Stigmas short and plumose. Filaments three. Anthers short, and cloven at either extremity. Length from twelve to twenty-four inches. Root annual and fibrous.

The specimen for illustration was collected in fields at Battersea, by Mr. Joseph Sidebotham, of Manchester.

SETARIA VERTICILLATA.   S. VIRIDIS.

SETARIA VERTICILLATA.

Beauvois. Hooker and Arnott. Parnell. Koch. Lindley. Babington. Kunth.

PLATE XXXI.—A.

Panicum verticillatum,	Linnæus. J. E. Smith. Knapp.
" "	Willdenow. Curtis. Graves.
" "	Schrader. Host. Ehrhart.
" "	Hull. Hudson. Withering.
" "	Reichenbach.
Pennisetum verticillatum,	Brown.
Gramen geniculatum,	Gerarde. Bauhin.

The Reflex Bristle-Grass.

Setaria—A bristle.   *Verticillata*—Whorled.

Setaria, *Beauvois.*—The Bristle-Grass, having a compound almost cylindrical spike, derives its name from the Latin *seta*, a bristle. There are three British species, yet perhaps better known under Linnæus's name of *Panicum.*

THE "Rough Panick-Grass" of "Smith's English Botany." Another doubtful British plant and interesting species, found only in cultivated fields in the counties of Middlesex and Norfolk, near London and Norwich.

Native of France, Italy, Germany, Switzerland, Holland, Belgium, North Africa, the United States of America, and Asia.

Stem erect, bearing four or five flat, rough, lanceolate leaves, with smooth striated sheaths, the upper one shorter than its leaf. Ligule short and blunt. Joints four. Inflorescence simple panicled. Branches brief. Rachis rough. Spikelets dorsally compressed, almost sessile, clustered, having involucral bristles longer than the spikelets, and each strongly dentate, the teeth pointing downwards. Glumes very unequal, two. Upper floret of two paleæ. Styles two. Stigmas short and plumose. Stamens three. Anthers deep purple in colour. Seeds shining and hard. Length from six to twenty-four inches. Root annual and fibrous.

The involucral bristles are much larger in *Setaria viridis*.

Flowers in July, and seeds ripen toward the close of September.

The specimen for illustration was gathered in Battersea Fields, by Mr. J. Sidebotham, of Manchester.

SETARIA VIRIDIS.

BEAUVOIS. HOOKER AND ARNOTT. PARNELL. KOCH. LINDLEY. BABINGTON. KUNTH.

PLATE XXXI.—B.

Panicum viride,	LINNÆUS. J. E. SMITH. KNAPP.
" "	SCHRADER. LEERS. WILLDENOW.
" "	CURTIS. GRAVES. EHRHART.
" "	HOST. HUDSON. WITHERING.
" "	HULL. REICHENBACH.
" crus-galli,	OEDER.

The Green Bristle-Grass.

Setaria—A bristle.   *Viridis*—Green.

AGAIN we have another dubious British Grass to describe, which is also an interesting species.

It is found in fields near London, Thetford, and Norwich.

Of no agricultural use. Small birds are fond of the abundant small seeds which it produces.

Native of France, Italy, Switzerland, Austria, Prussia, Spain, Portugal, Norway, Sweden Russia, North Africa, and the United States of America.

Grows on sandy soil on cultivated land.

Stem upright, bearing four or five flat, rough, lanceolate leaves, with smooth striated sheaths; upper one shorter than its leaf. Joints four. Inflorescence simple panicled; branches short; rachis hirsute. Spikelets dorsally compressed, crowded, almost sessile, having at the base long, rough, involucral bristles; dentate, teeth pointing upwards, and bristles above twice the length of the spikelet. Each spikelet of two glumes and two florets. Glumes very unequal. Styles two, long and smooth. Stigmas short and plumose. Stamens three. Anthers deep purple. Seeds smooth and hard. The involucral bristles in *Setaria verticillata* have the teeth pointing *downwards*. Length from three to eighteen inches. Root fibrous, annual.

Flowers in July, and seeds ripen at the end of September.

The specimen for illustration was gathered in Battersea Fields by Mr. Joseph Sidebotham, of Manchester.

SETARIA GLAUCA. Beauvois.

A THIRD species, *Setaria glauca*, has been discovered by Mr. Borrer, at Weybridge, in Surrey, and also at Hoddesdon, Hertfordshire, yet it has been considered a spurious British Grass. It has the dentations of the involucre erect, as in *S. viridis*, but differs from that species in having two glumellas, and in being wavy.

POA AQUATICA.
XXXII

POA AQUATICA.

LINNÆUS. J. E. SMITH. HOOKER AND ARNOTT. PARNELL. ABBOT.
GREVILLE. WILLDENOW. CURTIS. KNAPP. OEDER. LEERS.
SCHRADER. HOST. SIBTHORP. HUDSON. WITHERING. HULL. RELHAN.

PLATE XXXII.

Glyceria aquatica, SMITH. R. BROWN. BABINGTON.
Hydrochloa aquatica, LINDLEY.
Gramen aquaticum majus, RAY. GERARDE. LOBEL.

The Reed Meadow-Grass.

Poa—Grass.   *Aquatica*—Aquatic.

POA, *Linnæus.*—The Meadow Grass takes its name from the Greek, signifying *grass*, or *to feed*, on account of the whole genus being valuable for pasturage. There are fifteen British species.

A VERY common handsome Grass, growing in wet situations, such as undrained meadows, and along the edges of water. It is a valuable agricultural Grass for damp situations, as it can be mown twice a year, yields a heavy crop, and cattle are fond of its sugary flavour.

In Scotland it is found near Edinburgh, Forfar, Perth, and Dumbarton. In England, in the counties of Devon, Somerset, Kent, Surrey, Sussex, Middlesex, Norfolk, Suffolk, Cambridge, Oxford, Bedford, Worcester, Gloucester, Warwick, Nottingham, Leicester, York, Chester, Durham, Westmoreland, and Northumberland.

This summer I found it very abundant and large at Ambleside, in Westmoreland, and in many places round Nottingham.

Rare in Ireland.

Abroad it is a native of France, Italy, Germany, Norway, Russia, Sweden, and North America.

Stem erect, strong, smooth, striated, bearing seven or eight flat, lengthy, broad, rough-pointed leaves, with harsh sheaths, the upper one longer than its leaf, and having at its apex a brief, broad ligule. Joints smooth, six to eight in number. Inflorescence compound panicled. Panicle upright, large, branches rough, situated alternately in half-whorls on the rachis. Spikelets many, upright, having from four to eight florets, the upper ones large and ovate, the others smaller and narrower. Calyx consisting of two membranous, unequal-sized, blunt glumes, destitute of lateral ribs. Florets not webbed, consisting of two awnless paleæ. The exterior palea of lowest floret seven-ribbed. Stigmas compound and plumose. Styles larger than the stigmas. Length from three to six feet. Root perennial and creeping.

Flowers about the middle of July, and ripens its seed in a month.

The specimen for illustration was gathered near Gee-Cross, Cheshire, by Mr. Joseph Sidebotham, of Manchester.

POA DISTANS.

LINNÆUS. HOOKER AND ARNOTT.
J. E. SMITH. PARNELL. WILLDENOW. KNAPP. WITHERING. SCHRADER.
HOST. DEAKIN. LINDLEY. SINCLAIR.

PLATE XXXIII.

Sclerochloa distans,	BABINGTON.
Glyceria distans,	SMITH. KOCH. RALFS. REICHENBACH.
Festuca distans,	KUNTH.
Poa retroflexa,	CURTIS.
" *salina,*	POLLICH.
Aira aquatica, var.,	HUDSON.

The Reflexed Meadow-Grass.

Poa—Grass.   *Distans*—Distant.

A PRETTY Grass, but worthless to the agriculturist. Rare in Scotland. In England found in Devonshire, Somersetshire, Sussex, Cambridge, Bedford, Worcestershire, Leicestershire, Kent, Derbyshire, Nottinghamshire, York, Durham, and Northumberland. In Wales in Denbigh and Flintshire; and in Ireland near Dublin.

Abroad, a native of France, Italy, Germany, Switzerland, Prussia, Norway, and Sweden.

Stem upright, circular, polished, often decumbent at the base, having four flat acute leaves, with smooth striated sheaths; upper sheath longer than its leaf, having an obtuse ligule at the apex. Joints three, smooth. Inflorescence compound-panicled. Panicle upright. Branches rough and slender, arranged in twos, threes, or fives, the basal ones rigid, and bent downwards. Spikelets linear. Calyx consisting of two blunt, membranous, unequal glumes, three-ribbed; exterior glume half the length of inner glume. Florets of two equal-sized paleæ; exterior one of lowest floret five-ribbed; base in some degree hirsute; inner palea with two marginal ribs. Length from twelve to eighteen inches. Root perennial and fibrous.

Flowers at the beginning of July, and ripens its seed in a month.

Known from *Poa maritima* by the rough rachis and branches; from *P. procumbens* by the spreading branches of the panicle and the ribs of the glumes not being prominent; from *P. trivialis* by its smooth sheaths, blunt ligule, linear spikelets, and florets not being webbed; from *P. annua* in the roughness of the inner surface of the leaves and the branches of the panicle; and from *P. pratensis* in the linear spikelets, obtuse glumes, and in the florets not being webbed.

Variety *obtusa* found at Breeden, Leicestershire, by Mr. Babington, where it was growing in great quantities in the fields. The spikelets are larger, ligules shorter, and outer palea broader and more obtuse.

Variety *minor* also gathered at Breeden, by Mr. Babington. It is more dwarf, and the spikelets of fewer florets.

The specimen for illustration was gathered in the Nottingham Meadows.

POA FLUITANS.   P. MARITIMI.

POA FLUITANS.

SCOPOLI. HOOKER AND ARNOTT. PARNELL.
SMITH. GREVILLE. KNAPP. SALISBURY. SCHRADER. DEAKIN.

PLATE XXXIV.—A.

Festuca fluitans,	LINNÆUS. WILLDENOW. HUDSON. CURTIS.
" "	MARTYN. OEDER. HOST. SCHREBER.
" "	LEERS. HULL. SIBTHORP. ABBOT.
Glyceria fluitans,	SMITH. LINDLEY. RALFS. BROWN.
" "	SINCLAIR.
" *plicata,*	FRIES.
" *pedicellata,*	TOWNS.
Gramen fluviatile,	GERARDE.

The Floating Meadow-Grass.

Poa—Grass.   *Fluitans*—Floating.

THE present very graceful Grass will yield a good crop in damp meadows, and cattle are fond of its leaves. It usually grows in wet situations, in ditches, ponds, and along the margins of rivers.

Common throughout England, Scotland, and Ireland.

Native of France, Italy, Germany, Switzerland, Portugal, Spain, Norway, Sweden, New Holland, North America, and Northern Africa.

Stem, near the base, decumbent, the other portion upright, circular and smooth, bearing six or seven lengthy, linear, rough leaves, with rough striated sheaths, upper one extending beyond its leaf, and having a lengthy, ragged, pointed ligule at its apex. Joints smooth, about seven in number. Inflorescence simple-panicled. Panicle almost upright, long and slender, having rough branches arranged in pairs on the rachis. Spikelets few, long and linear, striped with green and white, of six to fourteen florets, the apex of basal floret stretching considerably beyond the large glume of the calyx. Calyx consisting of two obtuse, membranous, very unequal-sized glumes, destitute of lateral ribs. Florets of two paleæ; exterior palea of basal floret seven ribbed, dorsal rib not reaching to the apex, minutely dentate. Lateral ribs rough. Inner palea shorter, bifid, having two green marginal ribs.

Unlike all other of the *Poas*, and even *Festuca pratensis* of Hudson, which it most resembles, has only five ribs on outer palea, and the dorsal rib quite smooth.

Flowers towards the end of June, and ripens its seeds before the beginning of August.

The specimen for illustration was gathered at Ambleside.

POA MARITIMA.

Hudson. Hooker and Arnott. J. E. Smith. Parnell.
Linnæus. Knapp. Greville. Willdenow. Dickson. Schrader.
Roth. Oeder. Withering. Hull. Lightfoot.
Relhan. Deakin.

PLATE XXXIV.—B.

Sclerochloa maritima,   Lindley. Smith. Koch. Babington.
Festuca thalassia,   Kunth.
Glyceria maritima,   J. E. Smith. Ralfs. Reichenbach.

The Creeping Sea Meadow-Grass.

Poa—Grass.   *Maritima*—Maritime.

THE *Poa maritima* grows in salt-marshes, and is therefore not an agricultural Grass.

Found along the coast in the counties of Northumberland, Durham, Gloucester, Norfolk, Kent, Sussex, Somerset, and Devonshire. Also around Anglesea; more abundant along the coasts of Ireland and Scotland.

Abroad it is a native of France, Italy, Germany Norway, Sweden, Lapland, Iceland, and North America.

Stem upright, circular, and smooth, bent at the base. Each stem has three or four compressed, mostly folded, roughish leaves, with smooth swollen sheaths; upper sheath longer than its leaf, having a blunt decurrent ligule at the apex. Joints smooth, four in number. Inflorescence mostly simple-panicled, the panicle being upright, compact, and spreading when in flower, unilateral; rachis bare behind. Branches smooth, arranged in twos, threes, or fives, on the rachis. Spikelets linear, of six to ten florets; apex of basal floret stretches considerably beyond the large glume of the calyx. Calyx consisting of two membranous glumes, three-ribbed; inner glume nearly twice the length of outer glume. Florets of two equal-sized paleæ; exterior one of basal floret *sharp-pointed*, base slightly hirsute; smooth **above**, five-ribbed. Inner paleæ having two green marginal fringed ribs. Length six to twelve inches. Root creeping and perennial.

Flowers at the beginning of July, and the seeds become ripe in a month.

P*. *maritima differs from ***P*. *distans*** in having its rachis and branches **smooth to the** touch, its root creeping, and its central rib of outer **palea** extending to the summit. It also differs from ***P*. *procumbens*** in its smooth rachis and branches, and creeping root.

The specimen for illustration was gathered near Bristol, by Mr. **Joseph Sidebotham,** of Manchester.

POA PROCUMBENS.

POA PROCUMBENS.

CURTIS. SMITH. HOOKER AND ARNOTT. PARNELL. KNAPP. DEAKIN.

PLATE XXXV.—A.

Sclerochloa procumbens,   BEAUVOIS. LINDLEY. BABINGTON.
Glyceria procumbens,   SMITH. RALFS. REICHENBACH.
Poa rupestris,   WITHERING.

The Procumbent Sea Meadow-Grass.

Poa—Grass.   *Procumbens*—Procumbent.

ANOTHER sea-side Grass, growing on waste land, and more or less glaucous in appearance.

Frequent in Durham, Lancashire, Yorkshire, Gloucestershire, Norfolk, Sussex, Dorsetshire, and Devonshire. Rare in Ireland and Scotland. Also a native of France and Germany.

Stem somewhat prostrate, circular, and polished, having three flat, ribbed, sharp-pointed leaves, with smooth striated sheaths. Upper sheath extending considerably beyond its leaf, situated near the panicle, and having an oblong membranous ligule at its apex. Joints smooth, and three in number. Inflorescence both simple and compound-panicled. Panicle compact, lanceolate in shape, unilateral; rachis behind bare. Branches rough. Spikelets linear, mostly of five florets; apex of basal floret stretching some distance beyond the larger glume of the calyx. Calyx consisting of two unequal-sized, blunt, membranous glumes, with three bold ribs. Florets of two paleæ; exterior one of basal floret five-ribbed, slightly hirsute at the base, the centre rib stretching a little beyond the apex of the palea; inner palea somewhat shorter, and having two green marginal fringed ribs. Styles brief. Stigmas branched. Length from three to fifteen inches. Root annual.

Flowers about the middle of July, and ripens its seeds in a month.

Distinguished from *Poa maritima* in its annual fibrous root, in the rough rachis and branches, broad flat leaves, and in the greater length of the central rib of the outer palea. From *P. distans* in the compact panicle, the unilateral branches, in never being deflexed, and in the dorsal rib of outer palea stretching beyond the summit.

P. rigida, and also *P. loliacea* cannot be mistaken for it on account of their having acute glumes, destitute of lateral ribs.

The specimen for illustration was gathered at Blackpool, by Mr. Joseph Sidebotham, of Manchester.

POA RIGIDA.

LINNÆUS. HOOKER AND ARNOTT. SMITH. PARNELL.
GREVILLE. WILLDENOW. CURTIS. KNAPP. SCHRADER. HOST. HUDSON.
WITHERING. HULL. RELHAN. SIBTHORP. ABBOT. DEAKIN.

PLATE XXXV.—B.

Sclerochloa rigida,   BEAUVOIS. LINK. LINDLEY.
    "   "   BABINGTON.
Festuca rigida,   KUNTH. KOCH.
Glyceria rigida,   SMITH. RALFS. REICHENBACH.

The Hard Meadow-Grass.

Poa—Grass.   *Rigida*—Rigid.

THIS diminutive British Grass grows on walls, rocks, and in barren soil, and is therefore of no use to agriculturists.

A frequent Grass in England, Ireland, and south of Scotland.

Native of France, Italy, Germany, Switzerland, and Northern Africa.

Stem near the base decumbent, otherwise upright; circular, polished, and striated, having four or five linear, narrow, pointed, involute leaves, with smooth striated sheaths; upper sheath shorter than its leaf, and having at the apex a lengthy pointed ligule. Joints smooth, and three or four in number. Inflorescence simple-panicled. Panicle upright, short, lanceolate, rough, rigid, and unilateral. Spikelets linear, compressed, mostly consisting of seven florets; the apex of basal floret stretching slightly beyond the large glume of the calyx. Calyx of two acute unequal-sized glumes, destitute of lateral ribs. Florets of two paleæ; exterior one of basal floret five-ribbed, the marginal ribs being broad, and having a white line down the centre; intermediate ribs indistinct, dorsal rib dentate on the upper portion. Inner palea somewhat shorter, and having two green marginal ribs. Length from three to five inches. Root annual, fibrous, and woolly.

Flowers in the middle of July, and ripens its seed in a month.

Poa rigida is unlike all others except *P. loliacea;* however *P. rigida* has the apex of upper glume on a level with the base of the third floret, whilst in *P. loliacea* it is on a level with the base of the fourth floret.

The specimen illustrated was gathered on Colwick Park Wall, by Mr. Joseph Sidebotham.

POA PRATENSIS.
XXXVI

POA PRATENSIS.

LINNÆUS. HOOKER AND ARNOTT. PARNELL.
KOCH. SMITH. GREVILLE. RALFS. ABBOT. SIBTHORP. RELHAN.
HULL. WITHERING. HUDSON. SCHRADER. WILLDENOW.
CURTIS. KNAPP. DICKSON. SINCLAIR. BABINGTON. DEAKIN.

PLATE XXXVI.

Poa angustifolia,   LINNÆUS, (var. *subcærulea*, of HOOKER.)
" *subsærulea,*   SMITH.

The Smooth-Stalked Meadow-Grass.

Poa—Grass.   *Pratensis*—Of a meadow.

THE present common species, known from all other Grasses in having the lower florets webbed, is an early Grass, yielding a large crop, and liked by cattle. It is not, however, recommended to agriculturists on account of its creeping roots, which are calculated to impoverish the soil.

Native of England, Scotland, Ireland, France, Italy, Spain, Switzerland, Portugal, Prussia, Germany, Lapland, Norway, Sweden, Denmark, Iceland, United States, and Northern Asia.

Grows from the sea-level to three thousand feet altitude.

Stem upright, circular, polished; bearing three or four linear, flat, acute leaves, (edges rough,) with smooth striated sheaths. Upper sheath considerably longer than its leaf, having at its apex a blunt membranous ligule. Joints smooth. Inflorescence panicled, upright, and spreading, lower branches usually in threes or fives. Spikelets ovate, from three to five florets. Calyx of two almost equal-sized acute glumes; three-ribbed, the upper portion of the dorsal rib dentate. Florets of two awnless paleæ, the exterior palea of basal floret five-ribbed. Dorsal and marginal ribs hirsute on lower half, the base of the floret having a copious web suspending the calyx. Outer palea slightly longer than inner one; the latter sometimes bifid at summit, and having two marginal ribs. Length from twelve to fifteen inches. Root perennial and creeping.

Flowers at the commencement of June, and ripens its seed in a month.

This plant delights to grow in loose sandy soil; it is very subject to variety, nevertheless the webbed character of the florets, the upper leaf considerably shorter than its sheath, the obtuse ligule, and the hirsute marginal ribs are present in all the varieties.

Dr. Parnell enumerates the following:—

1.—*Planiculmis.* Stem stout; leaves short and broad, upper leaf folded. Common.

2.—*Umbrosa.* Tall and slender; panicle drooping; leaves long and narrow. Common in shady places.

3.—*Arida.* Stem sheathed nearly to summit; panicle drooping. Found in dry exposed situations.

4.—*Retroflexa.* Small and slender; lower branches of panicle suddenly bent downwards. Common under trees.

5.—*Muralis.* Slender and dwarf; panicle erect. Grows on shady walls.

6.—*Arenaria.* Stout and erect; large angular spikelets; outer palea seven-ribbed; glaucous. Found amongst sand on the sea coast.

The illustration is from a specimen from Highfield House.

v. COMPRESSA.

POA LOLIACEA.

Hudson. Hooker and Arnott. Parnell. Koch. Relhan.

PLATE XXXVII.—A.

Triticum loliaceum,	Smith. Hooker. Willdenow.
" "	Withering. Knapp. Schrader.
" "	Deakin. Reichenbach. Ralfs.
" *unilaterale,*	Aiton. Host. (*Not of* Linnæus.)
Catapodium unilaterale,	Link. Lindley.
Sclerochloa loliacea,	Woods. Babington.

The Dwarf Wheat Meadow-Grass.

Poa—Grass.   *Loliacea*—Made of Tares.

POA LOLIACEA grows on rocks and sandy soils along the sea coast in small tufts, and is a useless agricultural Grass.

Found in the counties of Cornwall, Devon, Dorset, Sussex, Somerset, Hants, Kent, Essex, Suffolk, Norfolk, Cambridge, York, Lancashire, Cumberland, Durham, and Northumberland. In Wales, in Flint, Glamorgan, and the Island of Anglesea. Frequent on the coast of Fife. Occasionally in Ireland.

Found also in France, Italy, Germany, Spain, and Portugal.

Stem ascending, slightly curved, stout, smooth, and striated, bearing three or four linear, smooth leaves, with smooth striated sheaths; upper one equal in length to its leaf, and having an obtuse, ragged ligule at its apex. Lower sheaths shorter than their leaves. Inflorescence racemed. Spikelets oblong-ovate, on brief, stout footstalks, arranged alternately on either side of the rough rachis, all in one direction, so as to hide the rachis and front, and to leave it bare behind. From eight to twelve florets. Calyx of two equal, acute glumes, destitute of lateral ribs. Dorsal rib prominent. Florets of two paleæ, exterior one of basal floret five-ribbed, marginal ribs broad, having a white band down the centre.

It is occasionally difficult to recognise *P. loliacea* from *P. rigida*.

Length from two to five inches. Root annual and fibrous.

Flowers in the middle of July, and ripens its seed in a month.

The specimen for illustration was gathered in the Island of Anglesea, by Mr. Joseph Sidebotham.

POA COMPRESSA.

LINNÆUS. HOOKER AND ARNOTT. SMITH. PARNELL. KOCH. LEERS.
SCHRADER. LINDLEY. WILLDENOW. KNAPP. HOST. OEDER.
DEAKIN. SINCLAIR. MACKEIGHT. KUNTH. BABINGTON. RALFS. RELHAN.
DILLWYN. SIBTHORP. DICKSON. WITHERING. HUDSON.

PLATE XXXVII.—B.

Poa subcompressa,   PARNELL.
" *polynoda,*   PARNELL.

The Flat-stemmed Meadow-Grass.

Poa—Grass.   *Compressa*—Compressed.

AN early Grass, growing well on poor soils and in dry stony places, but not productive, and therefore useless to agriculture.

Common in England, Scotland, and Ireland. Found in France, Italy, Switzerland, Germany, Prussia, Russia, Norway, Sweden, Iceland, Greenland, and North America.

Stem upright; base decumbent, much compressed, hence the name; bearing four or five somewhat short, flat, acute leaves, with rough edges and with smooth striated sheaths; upper sheath of same length as its leaf, and having a short obtuse ligule at the apex. Joints five, smooth. Inflorescence simple-panicled. Panicle somewhat unilateral, upright, compact, except when in flower, then spreading. Branches short, the basal ones distant. Spikelets ovate, compressed, and of five or seven florets. Calyx of two acute, about equal-sized glumes, frequently tinged with purple; three-ribbed, upper portion of central rib dentate. Florets of two paleæ, exterior one of basal floret three-ribbed; base furnished with a delicate web suspending the calyx. Inner palea having two green marginal ribs.

Length twelve inches; colour dark green. Root perennial and creeping.

Flowers in the middle of July, and ripens its seed in the middle of August.

The specimen for illustration was gathered at Bristol, by Mr. Joseph Sidebotham.

POA TRIVIALIS.
XXXVIII

POA TRIVIALIS.

Linnæus. Hooker and Arnott. Smith. Parnell. Koch.
Greville. Kunth. Lindley. Willdenow. Curtis.
Knapp. Sinclair. Schrader. Host. Deakin. Macreight. Babington.
Ralfs. Hudson. Withering. Hull. Relhan. Sibthorp. Abbot.

PLATE XXXVIII.

Poa dubia,	Leers.
" *scabra,*	Ehrhart.
" *setacea,*	Hudson.

The Roughish Meadow-Grass.

Poa—Grass.   *Trivialis*—Trivial.

A VALUABLE agricultural Grass for moist, rich, and sheltered soils.

Common throughout England, Scotland and Ireland. Found in France, Italy, Germany, Spain, Portugal, Prussia, Denmark, Switzerland, Sweden, Norway, Lapland, Iceland, Asia, and North America.

Stem upright; base decumbent, circular and rough, bearing five or six thin, flat, acute, rough leaves, with rough, striated sheaths; upper one much longer than its leaf, and having a lengthy pointed ligule at the apex. Joints smooth. Inflorescence panicled; panicle upright; branches spreading, rough, basal ones in threes or fives. Spikelets ovate, compressed, and of two to five awnless florets; apex of basal floret stretching slightly beyond the large glume of the calyx. Calyx of two equal acute glumes, upper one three-ribbed, lower one destitute of lateral ribs. Dorsal ribs strongly dentate. Florets of two paleæ; exterior one of basal floret five-ribbed, the base furnished with a delicate web suspending the calyx. Inner palea having two green, marginal, fringed ribs.

R

Length from twelve to eighteen inches. Root perennial and creeping.

Blooms in the third week in June, and ripens its seed in the middle of July.

There is a slender variety known as var. *parviflora*, with small spikelets of one or two florets, common in woods.

For specimens I am indebted to Dr. Wilson, of Nottingham; Miss Millet, of Penzance; and Mr. Joseph Sidebotham, of Manchester.

POA BULBOSA.

LINNÆUS. HOOKER AND ARNOTT. SMITH. PARNELL. BABINGTON.
KUNTH. KOCH. KNAPP. WITHERING. LINDLEY.
WILLDENOW. HUDSON. SCHRADER. DEAKIN. MACREIGHT. DILLWYN.
HOST. REICHENBACH. RALFS. HULL.

PLATE XXXIX.—A.

The Bulbous Meadow-Grass.

Poa—Grass.   *Bulbosa*—Bulbous.

A GRASS growing on the sandy shores of the south and east of England abundantly, especially in Norfolk and Suffolk. Of inferior agricultural merits.

Native of France, Italy, Spain, Portugal, Germany, Siberia, and North Africa.

Stem circular, smooth, hollow, and striated, bearing four or five flat, acute leaves, with smooth, striated sheaths, the upper sheath much longer than its leaf. Joints mostly three. Inflorescence panicled, branches rough. Spikelets ovate, green, or tinged with purple; composed of two glumes, and three or four florets. Glumes equal, and three-ribbed, keels above dentate. Florets longer than the glumes copiously webbed at the base, of two paleæ, exterior one of basal floret five-ribbed. Styles two. Stigmas feathery. Filaments three, and feathery. Anthers notched at either extremity.

Length from five to nine inches. Root perennial and bulbous; soon after flowering the leaves wither, after which the bulbs lie loose on the sand till autumn, when they again take root.

Flowers in April and May.

The specimen for illustration was gathered near Yarmouth by Mr. T. Coward.

POA ALPINA.

Linnæus. Hooker and Arnott. Smith. Parnell. Koch. Lindley.
Knapp. Willdenow. Lightfoot.
Schrader. Host. Wahlenberg. Deakin. Sinclair. Macreight.
Kunth. Babington. Ralfs.

PLATE XXXIX.—B.

Poa cæsia,   Smith.
" *glomerata,*   Don.

The Alpine Meadow-Grass.

Poa—Grass.   *Alpina*—Alpine.

AN early useless Grass, generally growing at from three to four thousand feet elevation.

In England, found in Yorkshire; Wales, Caernarvon; Scotland, Perth, Forfar, Aberdeen, and Inverness.

Native of France, Italy, Germany, Switzerland, Russia, Norway, Sweden, Lapland, Iceland, Greenland, and North America.

Stem upright, circular, smooth, bearing two or three short flat leaves, with smooth striated sheaths; upper sheath much longer than its leaf, and having a lengthy membranous ligule at the apex. Upper leaf folded, compressed, and rounded behind the apex. Joints two, smooth. Inflorescence panicled. Panicle compact and erect. Branches rough; basal ones in pairs. Spikelets broadly-ovate, commonly viviparous. Usually four awnless florets; summit of basal floret extending beyond the calyx. Calyx of two broad equal glumes, three-ribbed. Keels minutely dentate. Florets not webbed, of two paleæ; basal exterior one three-ribbed. Inner palea membranous, and rather shorter. Length from four to twelve inches. Root perennial, fibrous, and tufted.

Poa alpina may be distinguished from *P. laxa*, in having the panicle upright, the root tufted, and the upper leaf folded.

Blooms in the third week in May, and becomes ripe at the end of June.

The specimen for illustration was gathered on Snowdon, by Mr. Joseph Sidebotham, of Manchester.

POA NEMORALIS.   P. ANNUA.

POA NEMORALIS.

LINNÆUS. HOOKER AND ARNOTT. SMITH. PARNELL. GREVILLE. LEERS.
KOCH. WILLDENOW. KNAPP. SCHRADER. HOST. EHRHART.
OEDER. DEAKIN. SINCLAIR. LINDLEY. MACREIGHT. KUNTH. RALFS.
BABINGTON. WITHERING. HULL. RELHAN. SIBTHORP. ABBOT.

PLATE XL.—A.

Poa *glauca*,	SMITH. LINDLEY. SINCLAIR.
" *Parnelli*,	BABINGTON. PARNELL.
" *montana*,	PARNELL.
" *Balfourii*,	PARNELL.
" *angustifolia*,	HUDSON.
" *pratensis*,	WITHERING. HULL.

The Wood Meadow-Grass.

Poa—Grass.   *Nemoralis*—Of a wood.

A VALUABLE agricultural Grass usually growing in woods, yet succeeding well when cultivated in a field.

A common English Grass; abundant in Ireland, yet less frequent in Scotland.

Found also in Norway, Sweden, Denmark, Lapland, Iceland, France, Italy, Spain, Germany, Prussia, Northern Asia, and the United States.

Stem upright, slender, and compressed, bearing five or six linear, flat, narrow, acute leaves, with smooth striated sheaths, the upper sheath not longer than its leaf, and bearing at its apex a brief obtuse ligule. Joints about five, smooth. Inflorescence compound-panicled. Panicle somewhat pendulous, spreading near the base in twos, threes, or fours. Spikelets ovate, acute, of three or five awnless florets; the apex of basal one stretching slightly beyond the large glume of the calyx. Calyx of almost equal, acute, three-ribbed glumes, the dorsal rib being dentate above. Florets of two paleæ; the exterior one of basal floret five-ribbed. The calyx is suspended in a silky web at the base of the floret. Length from eighteen to twenty-four inches. Colour light green. Root perennial and creeping.

Variety *angustifolia*. Common. It has the first joint near the panicle, and the spikelets have only two florets.

Poa nemoralis is distinguished both from *P. trivialis* and *P. pratensis* in the upper sheath not extending beyond its leaf, and from *P. montana* and *P. polynoda* in the webbed florets.

Blooms in the third week in June, and ripens its seed at the close of July.

For specimens I am indebted to Dr. Wilson, of Nottingham, and to Mr. J. Sidebotham, of Manchester.

The specimen illustrated was gathered in Reddish Vale.

POA ANNUA.

LINNÆUS. HOOKER AND ARNOTT. SMITH. PARNELL. GREVILLE. KOCH.
WILLDENOW. CURTIS. MARTYN. STILLINGFLEET. KNAPP.
SCHRADER. HOST. LEERS. EHRHART. DEAKIN. SINCLAIR. LINDLEY.
KUNTH. MACREIGHT. BABINGTON. REICHENBACH. DILLWYN.
RALFS. HUDSON. WITHERING. HULL. RELHAN. SIBTHORP. ABBOT.

PLATE XL.—B.

The Annual Meadow-Grass.

Poa—Grass. *Annua*—An annual.

THE commonest of all Grasses, growing in any conceivable situation, and flowering throughout the summer. Found in all altitudes up to four thousand feet. Not a profitable agricultural Grass.

Found also in France, Italy, Switzerland, Germany, Spain, Portugal, Norway, Sweden, Denmark, Lapland, North Africa, North Asia, and in North and South America.

Stem ascending, most procumbent at the base, smooth, flattish, bearing four or five smooth, flat, flaccid, sword-shaped, vivid green leaves, often crumpled on the margins, with smooth compressed sheaths; upper sheath longer than its leaf, having a thin membranous acute ligule at the apex. Inflorescence compound-panicled. Panicle upright, outline triangular and spreading. Branches smooth, mostly in pairs. Spikelets ovate-oblong, mostly of five to eight awnless florets, commonly tinged with green, white, and purple; apex of basal floret stretching beyond the large glume of the calyx. Calyx of two unequal acute glumes, three-ribbed, dorsal rib dentate above. Florets of two paleæ, not webbed; exterior one of basal floret five-ribbed; ribs smooth. Inner palea membranous, shorter, having two green marginal ribs. Filaments three. Anthers brief, pendulous, and notched at either extremity. Styles two, short and naked. Stigmas feathery. Length from five to fifteen inches. Root creeping, throwing out fibres at the lower joints.

Poa annua differs from *P. pratensis* in the florets not being webbed. Blooms all summer through.

The specimen illustrated was gathered at Highfield House.

TRIODIA DECUMBENS.   DACTYLIS GLOMERATA.

TRIODIA DECUMBENS.

Beauvois. Hooker and Arnott. Parnell. Koch.
Lindley. Babington. Reichenbach. Deakin. Sinclair.
Smith. Ralfs.

PLATE XLI.—A.

Poa decumbens,	J. E. Smith. Hull. Withering.
" "	Greville. Hooker. Knapp.
" "	Schrader. Host.
Festuca decumbens,	Linnæus. Willdenow. Hudson.
" "	Oeder. Leers. Ehrhart.
" "	Dickson. Relhan. Abbot.
Danthonia decumbens,	De Candolle. Kunth. Macreight.
Melica decumbens,	Weber.

The Decumbent Heath Grass.

Triodia—Three teeth.   *Decumbens*—Decumbent.

Triodia.—*Brown.* Known as the Heath Grass. Only a solitary British example, the *Triodia decumbens*, which is common on our moors and mountains. The name is derived from the Greek.

A COMMON species, growing both on wet land and dry mountains, to an elevation of one thousand feet. Of but little agricultural interest.

Found in Scotland, England, Ireland, France, Italy, Switzerland, Turkey, Greece, Germany, Spain, Portugal, Prussia, Norway, and Sweden.

Stem circular, smooth, and striated, having three or four narrow linear leaves, with slightly hirsute sheaths, upper one shorter than its leaf, and bearing at its apex a tuft of hairs instead of a ligule. Leaves smooth, except near the apex, where very rough. Joints smooth. Inflorescence simple-panicled. Panicle consisting of few spikelets.

Spikelets of large size, consisting of four awnless florets, which do not extend beyond the calyx. Spikelets upright, having smooth footstalks, which are placed alternately on the rachis. Calyx of two smooth acute glumes, three-ribbed. Florets of two paleæ, exterior one of basal floret ovate, five-ribbed, three-toothed at the apex, base hirsute. Inner paleæ obtuse, broad, and having two green marginal ribs. Length from five to twelve inches. Root perennial and somewhat creeping.

Flowers at the end of July, and ripens its seed in the first week in August.

DACTYLIS GLOMERATA.

LINNÆUS. HOOKER AND ARNOTT. J. E. SMITH. GREVILLE. MACREIGHT.
KUNTH. RALFS. WITHERING. PARNELL. LINDLEY.
BABINGTON. REICHENBACH. DEAKIN. SINCLAIR. WILLDENOW.
RELHAN. MARTYN. KNAPP. SCHRADER.
SCHREBER. HOST. LEERS. OEDER. SIBTHORP. HUDSON.

PLATE XLI.—B.

Bromus glomeratus,   SCOPOLI.
Gramen asperum,   BAUHIN. RAY.

The Rough Cock's-foot Grass.

Dactylis—A finger.   *Glomerata*—In heaps.

DACTYLIS. *Linnæus.*—The Cock's-foot Grass. There is only a solitary British species. It is abundant everywhere. Name derived from the Greek.

THIS very common species is to be found everywhere. It is a rough harsh Grass, not liked by cattle, and where it predominates in a field, the produce, although increased in bulk, is rendered inferior in quality.

Native of England, Scotland, Ireland, Wales, Norway, Sweden, Denmark, Germany, Spain, Portugal, France, Russia, North Africa, and North America, growing to an altitude of one thousand feet above the sea.

Stem upright, circular, striated, and very rough, bearing five or six linear, flat, acute, widely-spreading, very rough (edges minutely toothed) leaves, with rough striated sheaths. Joints smooth and polished. Inflorescence compound-panicled. Panicle upright, tufted, and dense. Rachis and branches harsh. Spikelets crowded, unilateral, mostly of three florets. Calyx of two unequal glumes; hirsute. Florets of two paleæ; exterior one of basal floret longer than the calyx; five-ribbed.

Keel hirsute. Inner palea membranous. Length two to three feet. Root perennial, fibrous, and tufted.

Although a coarse Grass, still it must be looked upon as one of our most productive species, and when young, cattle will feed upon it readily. It will grow in almost any situation, from a wood to an open waste.

Dactylis glomerata continues flowering from June to August.

The specimen illustrated was gathered at Highfield House.

BRIZA MEDIA.
XLII

BRIZA MEDIA.

LINNÆUS. HOOKER AND ARNOTT. J. E. SMITH. PARNELL. KOCH. KUNTH. MACREIGHT. HUDSON. WITHERING. RELHAN. SIBTHORP. DICKSON. LINDLEY. GREVILLE. BABINGTON. REICHENBACH. DEAKIN. SINCLAIR. WILLDENOW. MARTYN. KNAPP. SCHRADER. HOST. LEERS. OEDER. RALFS.

PLATE XLII.

The Common Quaking Grass.

Briza—To droop.   *Media*—Ordinary.

BRIZA. *Linnæus.*—The Quaking Grass. There are two British species, one of which is abundant and the other rare. Name derived from the Greek.

AN exceedingly pretty Grass, very useful for decorative purposes, yet not a valuable agricultural Grass, owing to its not flourishing except in impoverished poor soil.

Common in poor land throughout England, Scotland, and Ireland. Found in France, Italy, Switzerland, Germany, Spain, Portugal, Prussia, Russia, Turkey, Greece, Norway, Sweden, and the United States.

It is found growing at an elevation of one thousand five hundred feet.

Stem upright, circular, smooth, and slender, bearing four flat acute leaves, with smooth striated sheaths, upper one extending considerably beyond its leaf, and having a short obtuse ligule at its summit. Joints smooth. Inflorescence simple or compound-panicled. Panicle broad, upright, and triangular. Branches exceedingly slender, spreading, and smooth, arranged in alternate pairs. Spikelets compressed, broadly ovate, purple, brown, and white, pendulous on long thin footstalks, and consisting of about seven awnless florets, extending beyond the calyx. Calyx of two broad obtuse glumes; three-ribbed. Florets of two paleæ, exterior one of basal floret compressed, broad, and obtuse. Base lobed, destitute of lateral ribs. Inner palea membranous, having two green marginal ribs. Length from twelve to eighteen inches. Root perennial, fibrous, and tufted.

Flowers at the end of June, and ripens its seed in July.

The specimen illustrated was gathered near Nottingham.

BRIZA MINOR.
XLIII

BRIZA MINOR.

LINNÆUS. HOOKER AND ARNOTT. J. E. SMITH. PARNELL.
BABINGTON. WITHERING. KOCH. KUNTH. REICHENBACH. DEAKIN.
WILLDENOW. SCHRADER. HOST.
HUDSON. HULL. DICKSON. MACREIGHT. RALFS.

PLATE XLIII.

Briza aspera,   KNAPP.

The Small Quaking Grass.

Briza—To droop.   *Minor*—Less.

AN exceedingly rare Grass, found near Bath, in Cornwall, and at Clifton, Nottinghamshire.

Native of Guernsey, Jersey, France, Italy, Switzerland, Spain, Portugal, Germany, Turkey, Greece, and Northern Africa.

It grows in dry sandy pastures.

Stem circular, smooth and hollow, carrying five or six flat, acute, roughish leaves, with smooth striated sheaths, upper one extending beyond its leaf. Joints five in number. Inflorescence compound-panicled, with roughish slender branches. Spikelets triangular, compressed, green, white, and purple in colour, consisting of two glumes and five or six florets. Glumes equal, broad, three-ribbed. Florets consisting of two unequal paleæ; exterior one of basal floret broad, gibbous behind, margin membranous, destitute of lateral ribs; inner palea flat, and having two broad green ribs. Styles two. Stigmas plumose. Filaments three. Anthers notched at either extremity.

Flowers in July, seeds ripen at the end of August.

The specimen illustrated was gathered at Penzance, by Mr. Joseph Sidebotham, of Manchester.

CYNOSURUS CRISTATUS.

LINNÆUS. J. E. SMITH. HOOKER AND ARNOTT. PARNELL. MACREIGHT.
KUNTH. RALFS. LINDLEY. GREVILLE. BABINGTON.
REICHENBACH. DEAKIN. SINCLAIR. RELHAN. WILLDENOW.
MARTYN. KNAPP. STILLINGFLEET.
SCHRADER. HOST. SCHREBER. LEERS. OEDER. WITHERING.

PLATE XLIV.—A.

Gramen cristatum.   BAUHIN. RAY. GERARDE.

The Crested Dog's-tail Grass.

Cynosurus—Dog's-tail.   *Cristatus*—Crested.

CYNOSURUS. *Linnæus.*—The Dog's-tail Grass. Two British species; one the *Cynosurus cristatus*, common; the other *C. echinatus*, local. Name derived from the Greek.

A VALUABLE permanent pasture Grass, but too dry and hard for hay. It flourishes best in clayey soils. Cattle are fond of the young leaves, yet reject the flower stalks, which cause the fields to look unsightly when they become dry. The stems are valuable for the manufacture of plait for Leghorn bonnets.

Common in England, Scotland, and Ireland.

Native of France, Italy, Germany, Switzerland, Portugal, Spain, Prussia, Norway, and Sweden, extending to an elevation of two thousand feet.

Stem circular, smooth, erect, and striated, having five flat, narrow, acute leaves, with smooth sheaths. Joints smooth. Inflorescence simple-panicled. Panicle upright, one inch and a half in length, linear, green when young, eventually brown. Spikelets consisting of three to five florets, having at the base a handsome pectinated involucre, which, together with the spikelets, point to one side of the rachis. Calyx of two narrow membranous glumes, destitute of lateral ribs. Florets of two paleæ; exterior one of basal floret extending beyond the calyx, form ovate-lanceolate, indistinctly five-ribbed; inner palea membranous, almost transparent. Length twelve to eighteen inches. Root perennial, fibrous, and tufted.

Flowers at the beginning of July, seeds become ripe in the middle of August.

The specimen illustrated was gathered at Highfield House.

CYNOSURUS ECHINATUS.

Linnæus. Hooker and Arnott. J. E. Smith. Parnell. Koth.
Lindley. Babington. Reichenbach. Deakin. Willdenow.
Sinclair. Knapp. Schrader. Host. Hall. Hudson. Withering.
Hull. Macreight. Kunth. Ralfs.

PLATE XLIV.—B.

The Tough Dog's-tail Grass.

Cynosurus—Dog's tail.   *Echinatus*—Covered with prickles; rough.

A LOCAL, curious, interesting Grass, found in Northumberland, Durham, Lancashire, Kent, and Sussex.

Native of the Shetland Isles, Jersey, France, Italy, Spain, Portugal, and North Africa.

Stem circular, smooth, upright, and minutely striated, carrying five flat, broad, tapering, rough leaves, with roughish sheaths; the upper sheath does not extend beyond its leaf, and is furnished at its apex with a lengthy pointed ligule. Joints smooth.

Inflorescence simple-panicled, crowded; colour silvery green. Panicle one inch in length, and half an inch wide, having brief rough branches, all inclined to one side. Spikelets of three awned florets, having at the base a pectinated involucre. Calyx of two equal-sized glumes, narrow, membranous, destitute of lateral ribs. Florets of two paleæ; exterior one of basal floret considerably shorter than the calyx, form ovate-lanceolate, five-ribbed, having a lengthy, slender, rough awn. Length from twelve to eighteen inches. Root annual and tufted.

Flowers the last week in June, and seeds ripen in August.

This species, which is of no agricultural value, is very distinct in appearance from the species last described, as will be seen by the illustration, as well as shewn by description.

The specimen illustrated was gathered at Hough-End, near Manchester, by Mr. Joseph Sidebotham.

FESTUCA PRATENSIS.
XLV.

FESTUCA PRATENSIS.

Hudson. Hooker and Arnott. J. E. Smith. Koch. Greville.
Kunth. Curtis. Martyn. Knapp. Schrader.
Relhan. Abbot. Babington. Ralfs. Macreight. Deakin.

PLATE XLV.

Festuca loliacea,	Smith. Hooker. Greville. Koch.
" *elatior,*	Linnæus. Host. Schreber. Leers.
" "	Ehrhart. Withering.
" *fluitans,* var.,	Hudson.
Bucetum loliaceum,	Parnell.
" *pratense,*	Parnell.
Schedonorus pratensis,	Lindley.

The Meadow Fescue Grass.

Festuca—........? *Pratensis*—Meadow.

Festuca. *Linnæus.*—The derivation of this word is dubious. It is a genus containing seven species according to Sir W. Hooker, but less in number according to Parnell, who separates several species under the name of *Bucetum*. The Grasses in *Festuca* have a loose panicle with many flowered spikelets, which are laterally compressed. Some of the species differ considerably from each other, as will be seen by reference to the figures and descriptions.

A VALUABLE Grass for agricultural purposes, growing on almost any soil, yielding a rich tender succulent hay, and being much liked by cattle.

Common throughout England, Scotland, Ireland, Germany, France, Switzerland, Italy, Russia, Norway, Sweden, Lapland, and the United States.

Stem upright, circular, smooth, and striated, bearing four or five lanceolate, acute, flat leaves, with smooth **striated sheaths, upper sheath** extending considerably beyond its leaf. **Joints four.** Inflorescence simple-panicled. Spikelets somewhat ovate, and of **five or six florets. Calyx** of two acute, unequal, smooth glumes, and three-ribbed. **Florets of two equal** paleæ, exterior one of basal floret **somewhat** longer **than the calyx;** five-ribbed and membranous. **Length from** fifteen to twenty-four inches. Root perennial **and fibrous.**

Flowers at the end of May, and seeds ripe at the end of July.

The illustration is from a specimen given to me by Dr. Wilson, of Nottingham.

FESTUCA ELATIOR.
XLVI

FESTUCA ELATIOR.

LINNÆUS. HOOKER AND ARNOTT. SMITH. GREVILLE. CURTIS.
SINCLAIR. DEAKIN. KUNTH. SCHRADER. HULL.
HUDSON. WITHERING. RELHAN. SIBTHORP. ABBOT. RALFS. MACREIGHT.

PLATE XLVI.

Festuca arundinacea,	SCHREBER. EHRHART. VILLARS.
" "	BABINGTON.
Bucetum elatius,	PARNELL.
Schedonorus elatior,	LINDLEY.
Bromus littoreus,	WILLDENOW. HOST. SINCLAIR.

The Tall Fescue Grass.

Festuca—........?   *Elatior*—Lofty, tall.

A VALUABLE agricultural Grass for moist or damp situations, being nutritive and very productive.

Common in England, Scotland, and Ireland; found in France, Italy, Germany, Switzerland, Norway, Sweden, Lapland, and North America.

Stem erect, circular, smooth, and striated, bearing five to six flattish, linear, acute leaves, with striated sheaths; upper sheath extending beyond its leaf, and having a short ligule at the apex. Inflorescence compound-panicled, the first four or five spikelets arising immediately from the rachis on brief stalks, the remainder on simple and compound branches. Panicle large and spreading, leaning to one side. Rachis and branches rough. Spikelets ovate-lanceolate, consisting of five or six slightly awned florets. Calyx of two unequal acute glumes, the exterior one destitute of lateral ribs, the interior one three-ribbed. Florets of two equal paleæ; exterior one of basal floret longer than the glumes, five-ribbed, the dorsal rib ending in a brief rough awn. Length from three to five feet. Root perennial, forming large tufts.

Dr. Parnell describes a variety, *variegatum*, in which the large spikelets are variegated with purple and white.

Flowers at the commencement of July, and seeds ripe in the middle of August.

The illustration is from a specimen gathered near Manchester by Mr. Joseph Sidebotham.

FESTUCA GIGANTEA.
XLVII

FESTUCA GIGANTEA.

VILLARS. HOOKER AND ARNOTT. SMITH. KOCH. LINDLEY. BABINGTON. KUNTH. MACREIGHT.

PLATE XLVII.

Bromus giganteus,	LINNÆUS. HOOKER. WILLDENOW.
" "	HUDSON. CURTIS. KNAPP.
" "	SCHRADER. SCHREBER. LINDLEY.
" "	DEAKIN. RALFS. ABBOT.
" "	SIBTHORP. HOST. LEERS.
" "	EHRHART. WEIGEL. WITHERING.
Bucetum giganteum,	PARNELL.
Festuca triflora,	SMITH.
Bromus triflorus,	LINNÆUS. WILLDENOW. OEDER.

The Tall Bearded Fescue Grass.

Festuca—........?   *Gigantea*—Gigantic.

GROWING in damp shady situations, and of but little agricultural value, as although there is an abundant produce, it is of but little nourishment for cattle.

Common in England, Scotland, and Ireland. Found also in France, Germany, Switzerland, Russia, Norway, Sweden, and Denmark.

Stem upright, circular, smooth, and striated, bearing five or six broad, lanceolate, flat, rough leaves, with striated sheaths; upper one longer than its leaf, and having at its apex a brief decurrent ligule. Joints five. Inflorescence simple or compound panicled, the lower ones being branched. Panicle large, loose, and leaning to one side. Spikelets ovate-lanceolate, mostly of five awned florets. Calyx of two unequal, acute, three-ribbed glumes. Florets of two equal paleæ; exterior one of basal floret longer than the calyx, and five-ribbed; the inner one having two green marginal ribs. Root perennial, fibrous, and somewhat creeping. Length from three to four feet.

Flowers towards the end of July, and ripens its seed at the end of August.

The illustration is from a specimen gathered by Mr. Wilson.

FESTUCA UNIGLUMIS.

SOLANDER. HOOKER AND ARNOTT. SMITH. PARNELL. KOCH.
KUNTH. WITHERING. BABINGTON.
KNAPP. DICKSON. RALFS. MACREIGHT. DEAKIN.

PLATE XLVIII.

Vulpia uniglumis, LINDLEY. DUMORT.
Lolium bromoides, HUDSON. WITHERING. HULL.
Stipes membranacea, LINNÆUS. MANT.

The Single-glumed Fescue Grass.

Festuca—........?   *Uniglumis*—Single-glumed.

THIS Grass, which has no agricultural merits, grows in arid sandy situations, chiefly in the immediate neighbourhood of the sea.

Found in Ireland and Anglesea. In Suffolk, Sussex, Essex, Dorset, and Devon. Abroad in France, Germany, Italy, and Switzerland.

Stem upright and slender, bearing three or four small, narrow, involute leaves, with smooth striated sheaths; upper sheath extending considerably beyond its leaf. Joints three. Inflorescence racemed, subsecund. Spikelets of two glumes and five or six florets. Glumes exceedingly unequal; inner one long and narrow; exterior one almost obsolete. Florets of two paleæ; exterior one of basal floret of same length as the large glume; five-ribbed, and ending in a lengthened rough awn. Styles two. Filaments three, capillary; stigmas plumose; anthers notched at either extremity. Root annual and fibrous. Length from ten to fifteen inches.

Flowers in June, and the seed becomes ripe in the middle of July.

The illustration is from a specimen gathered at Southport, by Mr. Joseph Sidebotham, of Manchester.

FESTUCA SYLVATICA.

VILLARS. HOOKER AND ARNOTT. SCHRADER. HOST. BABINGTON.
KUNZE. MACREIGHT.

PLATE XLIX.

Poa sylvatica,	POLLICH. PARNELL.
" *trinevata,*	EHRHART. SCHRADER. WILLDENOW.
" "	OEDER.
Festuca calamaria,	SMITH. HOOKER. KNAPP. WADE.
Schedonorus sylvaticus,	LINDLEY.

The Reed Fescue Grass.

Festuca—........?   *Sylvatica*—The wood.

A SOMEWHAT rare Grass, of which cattle are extremely fond. Found in damp woods. In England procured in Westmoreland, Worcester, and Sussex; occasionally in Scotland and Ireland.
Found in France and Germany.

Stem somewhat harsh, circular, slender, and erect; carrying three or four broad, flat, rough, ribbed, pale green leaves, with rough striated sheaths; upper one extending beyond its leaf, and having at its apex an obtuse membranous ligule; the other sheaths shorter than their leaves. Joints four, the upper two naked. Inflorescence compound-panicled. Panicle spreading, in some degree pendulous. Branches slender, and situated in pairs on the rachis. Spikelets many, small, of three awnless florets. Calyx of two narrow, acute, membranous glumes, destitute of lateral ribs. Florets of two equal-sized paleæ, exterior one of basal floret rough, acute, three-ribbed, the dorsal rib serrated. Root creeping, tufted, and perennial. Length from twenty-four to thirty-six inches.

Flowers in the middle of July.

The illustration is from a specimen gathered at Ambleside, by Mr. Joseph Sidebotham, of Manchester.

FESTUCA BROMOIDES. F. OVINA.

FESTUCA BROMOIDES.

Linnæus. Parnell. Hooker and Arnott. Smith. Greville.
Babington. Kunth. Macreight. Deakin.

PLATE L.—A.

Festuca Myurus,   Smith. (*Not of* Linnæus.)
"   *pseudo-myuru*,   Koch.
"   *sciuroides*,   Koch.
Vulpia bromoides,   Dumort. Lindley.

The Barren Fescue Grass.

Festuca—........?   *Bromoides*—Wild oats.

A NOT uncommon but useless agricultural Grass.
Found in England, Scotland, Ireland, France, Italy, Germany, Switzerland, Belgium, and Holland.

Stem upright, circular, smooth, and slender, carrying three or four short, very narrow, frequently involute leaves, with smooth striated sheaths, the upper one extending considerably beyond its leaf, and having a very brief ligule at its apex. Joints three, smooth. Inflorescence simple-panicled, long and slender. Spikelets erect, of five awned florets. Calyx of two exceedingly unequal acute glumes, the basal one destitute of lateral ribs, whilst the uppermost one is three-ribbed. Florets of two paleæ, the exterior one of basal floret five-ribbed, and of same length as the large glume, ending in a long slender awn. Inner palea lanceolate, having two green marginal ribs. Root annual and fibrous. Length from two to twenty-four inches.

Flowers in the middle of June, and seeds become ripe in the middle of July.

Variety *nana* grows in dry exposed localities, and is very stunted.
Variety *pseudo-myurus* somewhat common in corn-fields.

The illustration is from a specimen gathered at Southsea, by Mr. T. Coward.

Festuca ovina—Variety vivipara.

FESTUCA OVINA.

LINNÆUS. HOOKER AND ARNOTT. SMITH. PARNELL.
KOCH. LINDLEY. GREVILLE. WILLDENOW. MARTYN. KNAPP. HOST.
SCHRADER. LEERS. EHRHART. BABINGTON. KUNTH.
MACREIGHT. DEAKIN.

PLATE L.—B.

Festuca vivipara,	SMITH. KNAPP. DON. SINCLAIR.
" hirsuta,	HOST.
" cæsia,	SMITH.
" tenuifolia,	SIBTHORP. SCHRADER.
" duriuscula,	LINNÆUS. SMITH. PARNELL.
" "	VILLARS. WILLDENOW. KNAPP.
" "	SCHRADER. HOST. LEERS.
" "	DEAKIN. FRIES. SINCLAIR.
" rubra,	LINNÆUS. SMITH.
" "	WITHERING. DEAKIN.
" heterophylla,	HŒNKE. WILLDENOW.
" nemorum,	LEYSSER. ROTH. SCHRADER.
" dumetorum,	SMITH. LINNÆUS. WILLDENOW.
" "	OEDER. SINCLAIR.

The Sheep's Fescue Grass.

Festuca—........?   *Ovina*—Sheep.

THE present species is very subject to variety, and indeed some authors have divided *Festuca ovina* and varieties into several distinct species.

Common throughout England, Scotland, and Ireland. Native of Lapland, Norway, Sweden, Russia, Iceland, North America, Switzerland, Germany, Italy, France, Spain, Portugal, Siberia, and Greenland.

Stem angular and rough; erect, carrying three or four involute short rigid leaves, with rough sheaths, the upper one much longer than its leaf, and having a brief bilobed ligule at the apex. Joints two or three. Inflorescence simple-panicled. Panicle brief, compact, unilateral, and erect. Spikelets of six florets, having brief awns. Calyx consisting of two acute unequal-sized glumes, the upper one three-ribbed, and the lower one destitute of lateral ribs. Florets of two paleæ, the exterior one of basal floret five-ribbed, the interior one bifid, and having two green marginal ribs. Length from three inches to two feet. Root perennial and somewhat creeping.

Of the variety *hirsuta*, which is common in rocky situations, the glumes and florets are hairy.

Vivipara. A singular mountain variety, having the inner palea changed into a kind of leaf.*

Angustifolia. Abundant in the Highlands; slender, long, and narrow leaves.

Cæsia. Glaucous and altogether larger.

Duriuscula. Upper leaf flat, and larger in size.

Filiformis. A way-side Grass; tall, slender, and drooping.

Arenaria. Sandy situations near the sea; panicle and leaves short.

Humilis. An alpine variety. Slender, panicle narrow.

Rubra. The largest variety. Sandy sea-side situations.

Situation seems to be the chief cause of the great variety of this species.

A valuable agricultural Grass, especially for sheep; early and productive, though small in size.

Comes into flower in the middle of June, and ripens its seed in the middle of July.

The specimen from which the illustration is taken, was gathered at Langdale, Westmoreland, by Mr. Joseph Sidebotham, and the variety *vivipara*, which the engraving at page 154 represents, in Paterdale, Cumberland, by the same gentleman.

---

* See page 154 for a wood-cut illustration.

BROMUS ERECTUS.

BROMUS ERECTUS.

Hudson. Hooker and Arnott. Smith. Parnell.
Koch. Lindley. Dickson. Knapp. Sinclair. Schrader. Oeder.
Kunth. Babington. Macreight. Deakin.

PLATE LI.

Bromus agrestis,   Allioni. Host.
" *perennis,*   Villars.

The Upright Oat-Grass.

Bromus—Food.   *Erectus*—Upright.

Bromus. *Linnæus.*—The Brome-Grass, of which there are a dozen British species, has a lax panicle, with many-flowered laterally-compressed spikelets. The name is derived from the Greek, signifying *food;* hence the present word, which the Greeks used for one of the Oat-Grasses.

A LARGE-GROWING species, of but little agricultural value. In England found in Somerset, Sussex, Kent, Surrey, Norfolk, Cambridge, Oxford, Worcester, and Yorkshire. In the Island of Anglesea, and occasionally in Ireland and Scotland.

Found in Norway, Sweden, Germany, France, and Italy.

Stem circular and smooth, habit erect; bearing four or five linear, harsh, hairy, nearly erect leaves, with hairy sheaths, the upper one having at its apex a brief ragged ligule. Joints five. Inflorescence simple-panicled or racemed. Raceme upright and compact. Spikelets upright, consisting of eight or nine awned florets, tinged with brownish purple. Calyx of two equal-sized acute glumes; upper one three-ribbed, basal one destitute of lateral ribs. Florets of two paleæ, exterior one of basal floret a third longer than the small glume of the calyx; summit bifid and membranous; seven-ribbed; dorsal rib minutely dentate, and ending in a rough awn. Anthers of a deep saffron colour. Length from two to three feet. Root perennial and fibrous.

Flowers towards the end of June; seeds ripe in a month.

Variety *hirsutum* a hairy variety.

The specimen illustrated was gathered at Congleton, by Mr. E. Wilson.

BROMUS ASPER.
LII

BROMUS ASPER.

Linnæus. Hooker and Arnott. Smith. Parnell. Greville. Koch. Lindley. Willdenow. Martyn. Knapp. Schrader. Host. Ehrhart. Babington. Kunth. Macreight. Deakin.

PLATE LII.

Bromus ramosus,	Linnæus.
" *nemoralis,*	Hudson.
" *nemorosus,*	Villars.
" *hirsutus,*	Curtis.
" *montanus,*	Pollich. Retzius.

The Hairy Wood Bromus.

Bromus—Food.   *Asper*—Rough.

A COMMON, tall-growing, coarse Grass, found in damp shady woods, and of scarcely any agricultural merits.

Native of England, Scotland, Ireland, France, Italy, Germany, Switzerland, Norway, Sweden, and Russia.

Stem upright, circular, and somewhat rough, carrying four or five broad, flat, sharp-pointed, rough leaves, with striated hairy sheaths. Joints five. Inflorescence simple-panicled. Panicle weeping in habit. Spikelets one inch long, linear-lanceolate, of about eight awned glossy brownish purple florets. Calyx of two unequal acute glumes, the upper one longest and three-ribbed. Florets of two paleæ, exterior one of basal floret longer than the calyx, summit bifid, five-ribbed, the dorsal rib dentate, and ending in a long rough awn. Inner palea having two green marginal ribs. Root annual or biannual, and fibrous. Length from two to three feet.

Blooms at the end of July, and seeds ripen at the end of August.

The illustration is from a specimen forwarded by Dr. Wilson, of Nottingham.

BROMUS STERILIS.
LIII

BROMUS STERILIS.

Linnæus. Hooker and Arnott. Smith. Parnell. Greville.
Lindley. Host. Koch. Willdenow. Curtis. Martyn.
Knapp. Gerarde. Sinclair. Schrader. Leers. Ehrhart. Kunth.
Babington. Macreight. Deakin.

PLATE LIII.

Bromus grandiflorus,   Weigel.

The Barren Brome-Grass.

Bromus—Food.   *Sterilis*—Barren.

A COMMON, road-side, useless agricultural Grass, growing in dry shady situations.

Found in England, Scotland, Ireland, France, Germany, Italy, Norway, Sweden, Lapland, and Northern Africa.

Stem circular, rough, and striated, carrying four or five flat, linear, pubescent, acute leaves, with rough striated sheaths, the upper one of the same length as its leaf, and having a blunt ragged ligule at the apex. Joints five, naked. Inflorescence panicled, pale green in colour, and sometimes tinged with purple. Panicle spreading and drooping, having long, slender, rough branches. Spikelets mostly of eight awned florets, lanceolate and lengthy. Calyx of two unequal acute glumes, the upper one having three rough ribs, the lower one destitute of lateral ribs. Florets of two paleæ, the exterior one of basal floret longer than the calyx; margins membranous; summit bifid; seven-ribbed, the dorsal rib ending in a rough awn, which is longer than the palea. Inner palea shorter, and having two green marginal ribs. Length from one to two feet. Root annual and creeping.

Flowers towards the close of June, and the seeds become ripe in a month.

The illustration is from a specimen gathered at **Bredbury, in Cheshire, by** Mr. Joseph Sidebotham, of Manchester.

BROMUS SEGALINUS.
LIV

BROMUS SECALINUS.

SMITH. HOOKER AND ARNOTT. KOCH. LINDLEY. PARNELL.
HULL. WILLDENOW. KNAPP. SCHRADER. HOST. EHRHART. LEERS.
RELHAN. ABBOT. MACREIGHT. KUNTH. RALFS.

PLATE LIV.

Serrafalcus secalinus, BABINGTON.
Bromus velutinus, SMITH.
" *multiflorus*, SMITH.
" *polymorphus*, HUDSON. WITHERING.
" *vitiosus*, WEIGEL.

The Smooth Rye Brome-Grass.

Bromus—Food. *Secalinus*—Rye.

A SOMEWHAT common Grass, growing in corn-fields, and a useless somewhat troublesome weed.

Native of England, Scotland, Ireland, France, Italy, Norway, Germany, Sweden, and West Asia.

Stem upright, circular, smooth, and striated, carrying four or five flat, soft, linear, pointed leaves, with striated sheaths, the upper sheath having an obtuse, ragged, membranous ligule at its apex. Lower sheaths soft and hirsute. Joints five. Inflorescence racemed or simple-panicled. Panicle upright, branches harsh. Spikelets ovate, yellowish green, mostly of seven awned florets, the apex of the large glume being half-way between the apex and base of the second floret. Calyx consisting of two almost equal broad glumes, with membranous margins; upper half of the keel dentate. Inner glume seven-ribbed; outer glume, which is smaller, three-ribbed. Florets of two paleæ, exterior one of basal floret oval, seven-ribbed, the dorsal rib ending in a rough awn.

Inner palea linear oblong, having two green marginal ribs fringed with colourless hairs.

Length from eighteen to twenty-four inches. Root annual and fibrous.

Flowers in the first week, and becomes ripe in the last week in June.

There are two well-known varieties.

Variety *velutinus* having large spikelets of from ten to fifteen florets.

Variety *vulgaris* is frequently more than thirty-six inches in length.

The illustration is from a specimen gathered in Chorlton fields, near Manchester, by Mr. Joseph Sidebotham.

BROMUS COMMUTATUS.
LV

BROMUS COMMUTATUS.

Schrader. Koch. Parnell. H. Watson. Hooker and Arnott.

PLATE LV.

Serrafalcus commutatus.   Parlatore. Babington.

The Tumid Field Brome-Grass.

Bromus—Food.   *Commutatus*—Changed.

A SOMEWHAT common species, growing in corn-fields and on road-sides.

Stem upright, circular, smooth, and striated, carrying four or five flat, soft, sharp-pointed leaves, with striated sheaths, the upper sheath having an obtuse ragged ligule at its summit. Joints five. Inflorescence usually simple-panicled. Panicle when young upright, when more mature pendant. Branches rough. Spikelets linear-lanceolate, brownish purple, mostly of ten awned florets. Calyx consisting of two almost equal, broad acute glumes; margin membranous. Upper half of the keels dentate. Outer glume three-ribbed; inner glume seven-ribbed. Florets of two nearly equal-sized paleæ, the exterior one of basal floret oval, rough, glossy, and somewhat longer than the glumes; seven-ribbed. Inner palea linear-oblong, having two green marginal ribs fringed with white hairs. Stigmas plumose. Length from nineteen to thirty-six inches. Root fibrous and annual.

Bromus secalinus is more linear and longer.

It flowers in the middle of June, and ripens its seed at the commencement of July.

The specimen figured was gathered at York, by Mr. Joseph Sidebotham, of Manchester.

BROMUS ARVENSIS.

Koch. Smith. Lindley. Parnell.

PLATE LVI.

Serrafalcus arvensis,   Godron. Babington.

The Taper Field Brome-Grass.

Bromus—Food.   *Arvensis*—Field.

A RARE, and by some authorities considered a doubtful British species.

Found on the coast of Durham, at Hebden-Bridge, Yorkshire, Southampton Bay, Box Hill, and about Edinburgh.

Native of England, Scotland, Ireland, France, Italy, Norway, Germany, Sweden, Lapland, and Western Asia.

An early Grass, and useful for sheep.

Stem upright, circular, hard, bearing four or five narrow, flat, hairy leaves, with striated sheaths, which are shorter than their leaves. Joints four. Inflorescence simple-panicled; branches rough. Spikelets linear-lanceolate, mostly of seven awned florets, reddish brown in colour. Apex of large glume midway between the base of the glume and the apex of the second floret. Glumes unequal, margins membranous, keels rough. Inner glume largest, and five-ribbed; outer glume three-ribbed. Florets of two paleæ, exterior one of basal floret longer than the glumes, summit bifid or entire; margins membranous. Inner palea thin, acute, white, membranous, and having two green ribs fringed with colourless hairs. Awns upright and rough. Styles two, and short. Stigmas plumose. Filaments three. Anthers lengthy and notched at either extremity. Length from ten to eighteen inches. Root annual and fibrous.

Flowers in June and July, and ripens its seed in the second week of August.

The illustration is from a specimen gathered at Hebden-Bridge, by Mr. Joseph Sidebotham.

BROMUS MOLLIS.   B. DIANDRUS.

BROMUS MOLLIS.

Linnæus. Hooker and Arnott. H. Watson. Parnell. Curtis.
Willdenow. Martyn. Knapp. Sinclair.
Schrader. Host. Schreber. Leers. Ehrhart. Weigel. Lindley.
Koch. Greville. Hull. Relhan.
Sibthorp. Abbot. Macreight. Kunth. Ralfs. Deakin.

PLATE LVII.—A.

Serrafalcus mollis,   Parlatore. Babington.
Bromus polymorphus,   Hudson. Withering.
" *hordeaceus,*   Linnæus.
" *nanus,*   Weigel.

The Soft Brome-Grass.

Bromus—Food.   *Mollis*—Soft.

A USELESS species, growing on poor land. Common in England, Scotland, and Ireland. Found also in France, Italy, Germany, Switzerland, Norway, Sweden, Denmark, North Africa and North America.

Stem hairy, upright, and circular, carrying three or four flat, linear-lanceolate, striated, hairy leaves, with striated sheaths, having a small obtuse jagged ligule. Joints four or five, in some degree hirsute. Inflorescence racemed or simple-panicled. Raceme upright; branches rough and hirsute, basal ones mostly in threes. Spikelets upright, ovate in form, deep green in colour, and mostly of ten awned florets. The apex of large glume midway between its base and the apex of the third floret. Calyx of two broad hirsute glumes; upper one seven-ribbed, lower ones five-ribbed, ending in a rough awn, mostly bifid at the summit.

Length from twelve to eighteen inches. Root annual and fibrous.

Blooms in the last week of May, and ripens its seed in the middle of June.

BROMUS DIANDRUS.

Curtis. Parnell. Hooker and Arnott. Smith.
Knapp. Graves. Sinclair. Babington. Reichenbach.
Ralfs. Deakin.

PLATE LVII.—B.

Bromus Madritensis,	Linnæus. Willdenow.
" "	Schrader. Host. Koch.
" "	Withering. Macreight.
" "	Kunth.
" gynandrus,	Roth.
" rigidus,	Roth.
" muralis,	Hudson.
" ciliatus,	Hudson.
Festuca Madritensis,	Desfontaines.

The Upright Annual Brome-Grass.

Bromus—Food.   Diandrus—........?

A RARE species, growing chiefly on rocks and walls, and on dry soils.

In England it occurs in the counties of Devon, Somerset, Gloucester, Hampshire, Surrey, Kent, Worcester, and Durham. In Scotland it has been found on the Fifeshire coast, and near Edinburgh. Abroad it occurs in France, Italy, Germany, and Switzerland.

Stem upright, circular, and polished, having three or four flat, linear, acute, hairy leaves, with striated sheaths. Upper sheath downy, having a short, blunt, ragged ligule; lower sheath hairy, the hairs pointing downwards. Joints four, smooth. Inflorescence racemed. Raceme upright and compact. Spikelets commencing direct from the rachis, on short footstalks, the basal ones mostly in twos or threes; generally of eight awned florets, brownish purple in colour. Calyx of two unequal acute glumes. Upper one three-ribbed, lower one destitute of lateral ribs. Florets of two paleæ, the exterior one of basal floret bifid, membranous, and extending beyond the calyx, seven-ribbed, the two marginal ribs on either side approximate, central rib dentate. Length from six to twelve inches. Root annual and fibrous.

Flowers towards the end of June, and ripens its seed at the end of July.

The specimen illustrated was gathered at Bristol, by Mr. Joseph Sidebotham, of Manchester.

BROMUS MAXIMUS.
LVIII

BROMUS MAXIMUS.

DESFONTAINES. SMITH.
HOOKER AND ARNOTT. PARNELL. KUNTH. BABINGTON. RALFS.

PLATE LVIII.

The Great Brome-Grass.

Bromus—Food.   *Maximus*—Great.

A RARE British species, of no agricultural value. Found in Jersey, France, Spain, and Africa.

Stem upright, circular, and hollow, having four or five flat, acute, downy leaves, with rough margins, and with striated sheaths, the upper one extending slightly beyond its leaf, and having a conspicuous ragged ligule. Joints four, usually naked. Inflorescence racemed, and upright in habit. The footstalks and rachis downy. The form of the spikelets lanceolate; length an inch and a quarter, and having awns an inch and a quarter in length; consisting of eight awned florets, and two unequal, lanceolate glumes. Florets of two paleæ, the exterior one of basal floret exceedingly rough and lanceolate, and having seven conspicuous rough ribs. Awns straight and rough. Styles two. Stigmas plumose. Filaments three, and anthers notched at either extremity. Length from twelve to twenty-four inches. Root annual and fibrous.

It is known from *B. sterilis* by the soft downy footstalks.

Comes into flower in the middle of June.

AVENA FATUA.
LIX

AVENA FATUA.

LINNÆUS. SMITH. HOOKER AND ARNOTT. PARNELL. KNAPP.
WILLDENOW. MARTYN. DON. SCHRADER.
HOST. LEERS. EHRHART. SCHREBER. KOCH. LINDLEY. WITHERING.
HUDSON. KUNTH. HULL. RELHAN.
SIBTHORP. ABBOT. WINCH. MACREIGHT. BABINGTON. DEAKIN. RALFS.

PLATE LI.

The Wild Oat-Grass.

Avena—Oat. Fatua—Wild.

AVENA. *Linnæus.*—The Oat-Grass has a lax panicle and laterally compressed spikelets. Awns long and twisted. In this family is the *Avena sativa*, or Cultivated Oat, an introduced species. Amongst our indigenous species are *Avena fatua, A. strigosa, A. pratensis, A. pubescens, A. flavescens,* and *A. planiculmis;* the latter has only been collected by one botanist, Mr. Murray, who discovered it at Glen Sannox, in the Isle of Arran.

THE Wild Oat-Grass is a common species in England and Ireland, yet much rarer in Scotland. It chiefly grows in corn-fields, and is a troublesome weed.

The awns, from their extreme sensitiveness to the moisture of the air, are manufactured into Hygrometers. The florets are also occasionally used as artificial flies for trout-fishing.

Native of France, Italy, Germany, Norway, Sweden, Lapland, Asia, and Northern Africa.

Stem upright, circular, and polished, having four or five flat, linear, rough, minutely-ribbed leaves, with smooth striated sheaths. Joints smooth. Inflorescence simple-panicled. Panicle spreading, and of large size. Rachis smooth, branches rough. Spikelets ample, pendulous, ovate-lanceolate in form; of two (sometimes three) florets. Calyx of two equal-sized, smooth, membranous, acute glumes, the exterior one seven-ribbed, the inner one eleven-ribbed. Florets of two paleæ, exterior one of basal floret ovate in shape, acute, eight-ribbed, and considerably shorter than the calyx. Awn above double the length of the floret, twisted and bent, and of a dull reddish colour. Length thirty-six inches. Root annual and fibrous.

Flowers at the commencement of July, and seeds become ripe at the end of August.

The illustration is from a specimen gathered at Congleton, by Mr. Joseph Sidebotham, of Manchester.

AVENA PRATENSIS.   A. PUBESCENS.

AVENA PRATENSIS.

LINNÆUS. HOOKER AND ARNOTT. SMITH. WILLDENOW.
KNAPP. HUDSON. HULL. SINCLAIR. SCHRADER.
HOST. LEERS. KOCH. WITHERING. RELHAN. SIBTHORP. RALFS
KUNTH. ABBOT. LIGHTFOOT. MACREIGHT. BABINGTON.
LINDLEY. DEAKIN.

PLATE LX.—A.

Trisetum pratense,	PARNELL.
Avena alpina,	SMITH.
" *planiculmis,*	SMITH. HOOKER.
" *bromoides,*	LINNÆUS. WILLDENOW. GOUAN.

The Narrow-leaved Perennial Oat-Grass.

Avena—Oat.   *Pratensis*—A meadow.

A COMMON species, of scarcely any agricultural merit.
Found in England, Scotland, Ireland, France, Italy, Germany, Spain, Portugal, Norway, Sweden, and Lapland.

Stem upright, almost circular, smooth, and minutely striated, having three or four linear-acute harsh leaves, with striated sheaths, upper sheath twice the length of its leaf, rough, and having a lengthy, narrow, membranous ligule; lower sheaths smooth, and shorter than their leaves. Joints three, smooth. Inflorescence compound-racemed, or simple-panicled. Panicle upright, lengthy, and compact. Rachis and branches rough to the touch. Spikelets large, oval in form, of four or five awned florets, and of the same length as the calyx. Calyx consisting of two acute unequal glumes, three-ribbed, having rough keels, and on the lower portion purplish. Floret of two paleæ, the basal exterior one frequently bifid, five-ribbed, base hirsute. Inner palea shorter, and having minutely fringed margins. Awns rough, and twisted at the base. Length from eighteen to twenty-four inches. Root perennial and fibrous.

Flowers at the commencement of June, and ripens its seed in six weeks.

There are two varieties. *Longifolium*, having lengthy, linear, flat leaves; growing in damp shady woods near the sea in the neighbourhood of Edinburgh. *Latifolium*, with short broad leaves; a tall stout plant.

The specimen illustrated was gathered at Tadcaster, by Mr. Joseph Sidebotham, of Manchester.

AVENA PUBESCENS.

LINNÆUS. SMITH. WILLDENOW. KNAPP. HOOKER AND ARNOTT.
SINCLAIR. SCHRADER. LEERS. HOST. OEDER.
EHRHART. KOCH. GREVILLE. HUDSON. WITHERING. HULL. RELHAN.
KUNTH. SIBTHORP. ABBOT. DEAKIN.
LIGHTFOOT. MACREIGHT. BABINGTON. REICHENBACH. RALFS.

PLATE LX.—B.

Trisetum pubescens,   PERSOON. PARNELL. LINDLEY.
Avena sesquitertia,   LINNÆUS.

The Downy Oat-Grass.

Avena—Oat.   *Pubescent*—Downy.

A GRASS deserving the attention of agriculturists, giving a good yield, and requiring but little nourishment from the soil.

A frequent Grass in England, Scotland, and Ireland. Native also of France, Italy, Germany, Russia, Norway, and Sweden.

Stem upright, circular, and smooth, and minutely striated, having about five soft, broad, flat, hairy leaves, with the upper sheath more than three times the length of its leaf, and having a conspicuous membranous ligule; lower sheaths not so long as their leaves. Joints three or four. Inflorescence compound-racemed or simple-panicled. The basal spikelets situated on lateral branches, whilst those near the apex are on brief footstalks. Panicle upright. Calyx of two unequal, membranous, acute glumes, the basal one destitute of lateral ribs, and shorter than the upper one. Florets of two paleæ, the exterior one of basal floret membranous on the upper portion; five-ribbed; colour reddish purple; base hirsute. Inner palea much shorter, and exceedingly thin. Length from twelve to twenty-four inches. Root perennial and creeping.

Flowers in the middle of June, and ripens it seed in the middle of July.

This species usually grows in dry situations in chalky or limestone districts.

The illustration is from a specimen gathered at Burton, by Mr. Joseph Sidebotham, of Manchester.

AVENA STRIGOSA.

Schreber. Hooker and Arnott. Smith. Parnell. Koch. Lindley.
Willdenow. Knapp. Don. Schrader. Host. Ehrhart.
Retz. Withering. Hull. Babington. Macreight. Kuth. Deakin.

PLATE LXI.

The Bristle-pointed Oat-Grass.

Avena—Oat.   *Strigosa*—Slender.

A COMMON species, growing in corn-fields, and differing from *Avena fatua* and *A. sativa* in having the florets ending in two long bristles.

Found in the counties of Notts., York, Durham, Sussex, Cornwall, and Denbigh. In Scotland, in Inverness, Aberdeen, Forfar, and Perthshire. The Island of Anglesea. Central Europe.

Stem upright, circular, and polished, bearing four or five somewhat broad, acute, glaucous, rough leaves, with smooth striated sheaths, the upper one extending beyond its leaf, and having an oblong membranous ligule at its apex. Joints smooth. Inflorescence simple panicled. Panicle inclined to one side, having rough lengthy lateral branches. Spikelets large and oval, of two awned florets. Calyx of two acute, smooth, membranous, somewhat unequal glumes, the basal one smallest and seven-ribbed, the other nine-ribbed. Ribs prominent and green. Florets of two paleæ, the exterior one of basal floret of same length as large glume, ending in two rough bristles; six-ribbed and rough. Inner palea linear, membranous, and shorter. Awn rough, thick, and bent. Length thirty-six inches. Root annual and fibrous.

Flowers at the commencement of July, and ripens its seed in six weeks.

A. strigosa much resembles the *A. sativa*, (the cultivated Oat,) but is known from it in the florets ending in two lengthy straight bristles.

The specimen illustrated was gathered at Highfield House, Nottinghamshire.

AVENA FLAVESCENS.   HORDEUM SYLVATICUM.

AVENA FLAVESCENS.

LINNÆUS. HOOKER AND ARNOTT. SMITH. KOCH. GEEVILLE. WILLDENOW.
CURTIS. KNAPP. SINCLAIR. SCHRADER.
HOST. SCHREBER. LEERS. EHRHART. WITHERING. HUDSON. HULL.
RELHAN. SIBTHORP. ABBOT. REICHENBACH. DEAKIN.

PLATE XLII.—A.

Trisetum flavescens,   BEAVEAUX. PARNELL. LINDLEY.
    "   "   BABINGTON. MACREIGHT. KUNTH.

The Yellow Oat-Grass.

Avena—Oat.   *Flavescens*—Yellow.

A FREQUENT species, found in dry meadows and pastures, in England, Scotland, and Ireland, France, Italy, Germany, Spain, Portugal, Norway, Sweden, Russia, and North Africa.

Sheep are very fond of this Grass.

Stem upright, circular, and polished, carrying six or seven flat, roughish, acute leaves, with striated sheaths, the upper one double the length of its leaf, and having a brief ligule at its apex. Joints smooth. Inflorescence panicled, the panicle being upright and spreading. The lower branches usually in fives. Spikelets numerous, upright, and diminutive, mostly of three-awned florets, which extend beyond the calyx. Calyx of two acute unequal membranous glumes, the upper glume being the largest and three-ribbed. Florets of two paleæ, exterior one of basal floret membranous. Apex bifid; base hirsute; five-ribbed. Inner paleæ linear, acute, and membranous. Awn twisted at the base, rough, and longer than the palea. Length from twelve to twenty-four inches. Root perennial and creeping.

Flowers in the middle of July, and ripens its seed in a month.

The specimen illustrated was gathered at Highfield House, Nottinghamshire.

HORDEUM SYLVATICUM.

Hudson. Hooker and Arnott. Babington. Parnell. Knapp. Martyn. Abbot. Deakin.

PLATE LXII.—B.

Elymus Europeus,	Linnæus. Smith. Hooker.
" "	Lindley. Koch. Willdenow.
" "	Withering. Schrader. Host.
" "	Ehrhart. Hull. Sibthorp.
" "	Kunth. Reichenbach.

The Wood Barley.

Hordeum—........? Sylvaticum—Wood.

Hordeum. *Linnæus.*—Spikelets in threes from the same joints of the rachis. There are four British examples, all being known by the form of their glumes. The Barley Grasses, under which designation these species are known, are rare in Scotland. The name is of dubious origin.

OCCURRING more especially in a chalky soil in woods in the counties of Derby, York, Northumberland, Bucks., Herts., Hunts., Wilts., Bedford, Oxford, and Denbigh. Native of France, Germany, Italy, Switzerland, Norway, and Sweden.

Of no agricultural value.

Stem upright, circular, somewhat smooth, bearing four or five lanceolate, rough, flat, pointed leaves, with rough striated sheaths, upper one extending beyond its leaf. Joints four. Inflorescence spiked, compact, three inches long. Rachis rough, angular and dentate. Spikelets in threes. Glumes three-ribbed, rough, equal, ending in a long rough awn. Floret of two paleæ, exterior one awned, rough, and five-ribbed; base hirsute. Inner palea two-ribbed and of same length. Awn of exterior palea rough, and commencing at the apex. Ovarium hirsute. Styles two, brief. Stigmas plumose. Filaments three. Anthers lengthy, and cloven at either extremity. Length twenty-four inches. Root perennial and fibrous.

Flowers in June, and ripens its seed the second week in August.

The illustration is from a specimen gathered at Cottril Clough, by Mr. Joseph Sidebotham, of Manchester.

HORDEUM PRATENSE.

HUDSON. HOOKER AND ARNOTT. SMITH. KUNTH.
DEAKIN. PARNELL. LINDLEY. RELHAN. MARTYN. KNAPP. SINCLAIR.
MACREIGHT. SCHRADER. EHRHART. WITHERING. SIBTHORP.
BABINGTON. REICHENBACH.

PLATE LXIII—A.

Hordeum nodosum,	KOCH. LINNÆUS.
" *secalinum,*	WILLDENOW. HOST.
" *maritimum,*	OEDER.
Gramen secalinum,	GERARDE. RAY.

The Meadow Barley.

Hordeum—........?   *Pratense*—A field.

FOUND in moist meadows and pastures in the counties of Somerset, Sussex, Kent, Surrey, Norfolk, Suffolk, Cambridge, Bedford, Oxford, Leicester, Worcester, Warwick, Nottinghamshire, Derbyshire, Cheshire, Durham, Northumberland, Flint, and Denbigh. In Scotland rare—near Edinburgh; occasionally in Ireland. Extending into central Europe.

An early species, and although common in Norfolk pastures is not considered a profitable agricultural Grass.

Stem circular, smooth, upright, and polished, carrying four or five linear, flat, somewhat hirsute leaves, with smooth striated sheaths; the upper one being longer than its leaf, and having a very brief ligule at its apex. Joints smooth. Inflorescence spiked. Spikes dense, and an inch and a half long. Rachis dentate. Spikelets in threes on each tooth of the rachis. Calyx of central spikelet consisting of two equal-sized glumes. Central floret of two paleæ; exterior one three-ribbed, and ending in a lengthy rough awn; inner palea acute, and only half the length. Length eighteen to twenty-four inches. Root perennial and fibrous.

Flowers at the commencement of July, the seeds becoming ripe in a month.

The specimen illustrated was gathered at Bristol, by Mr. Joseph Sidebotham, of Manchester.

HORDEUM MURINUM.

LINNÆUS. HOOKER AND ARNOTT. KUNTH. BABINGTON. KOCH. SMITH. PARNELL. GREVILLE. LINDLEY. WILLDENOW. REICHENBACH. RELHAN. CURTIS. MARTYN. KNAPP. HOST SINCLAIR. SCHRADER. DICKSON. SIBTHORP. ABBOT. OEDER. EHRHART. HUDSON. WITHERING. HULL. MACREIGHT. DEAKIN.

PLATE LXIII.—B.

Hordeum spurium,　　　　　GERARDE.

The Wall Barley.

Hordeum—........?　　　　*Murinum*—A wall.

A VERY common English Grass, growing on waste grounds, on roadsides, and by walls. Common in the south of Europe and in Germany.

A useless agricultural Grass.

Stem circular, upright, and smooth, carrying three or four linear, flat, somewhat hirsute roughish leaves, with smooth striated sheaths, the upper one extending beyond its leaf, and having at its apex a brief jagged ligule. Joints smooth. Inflorescence spiked; the spike two inches long, dense, and compact. Spikelets in threes, and consisting of one awned floret. Calyx of two equal-sized glumes, and ending in a lengthy rough awn. Central floret consisting of two paleæ, exterior one ovate and three-ribbed, interior one membranous. Length from twelve to thirteen inches. Root annual and fibrous.

Flowers at the commencement of July, and ripens its seed in a month.

The specimen illustrated was gathered in Nottingham Park.

HORDEUM MARITIMUM.   LXIV   TRITICUM JUNCEUM.

HORDEUM MARITIMUM.

WITHERING. HOOKER AND ARNOTT. DEAKIN. KUNTH. SMITH.
PARNELL. KOCH. LINDLEY. KNAPP.
MARTYN. SCHRADER. MACREIGHT. VALL. HOST. POURRET. HULL.
RELHAN. BABINGTON. REICHENBACH.

PLATE LXIV.—A.

Hordeum marinum,   HUDSON. DICKSON.
    "   *geniculatum*,   ALLIONI.
    "   *rigidum*,   ROTH.

The Sea-side Barley.

Hordeum—... ...?   *Maritimum*—Sea.

GROWING near the sea, on light dry sandy ground. Occurring in Devon, Somerset, Dorset, Sussex, Essex, Kent, Suffolk, Norfolk, Gloucester, Glamorgan, York, Durham, and Northumberland; and in Argusshire, where rare. Found also along the Mediterranean Sea, extending to the Baltic.

An injurious agricultural Grass.

Stem upright, circular, and polished, (base trailing) bearing four or five brief, narrow, rough, hirsute leaves, with smooth, striated sheaths, having a brief, membranous ligule at its apex. Joints smooth. Inflorescence spiked, uniform, and an inch in length. Rachis jointed and dentate. Spikelets in threes on either side of the rachis, and of one awned floret. Calyx of two equal-sized rough glumes. Floret of two paleæ, the exterior one ending in a lengthy rough awn; inner one half the length. Floret imperfect. Length from three to nine inches. Root annual and fibrous.

Flowers in the middle of June.

The specimen illustrated was gathered near Bristol, by Mr. Joseph Sidebotham, of Manchester.

TRITICUM JUNCEUM.

Linnæus. Hooker and Arnott. Koch. Withering.
Smith. Parnell. Lindley. Greville. Willdenow. Hull.
Knapp. Dickson. Schrader. Host. Oeder. Hudson. Babington.
Reichenbach. Kunth. Mackeight. Deakin.

PLATE LXIV.—B.

Agropyrum junceum,   Lindley. Beauvais.

The Rushy Sea Wheat-Grass.

Triticum—Rubbed.   *Junceum*—A rush.

Triticum. *Linnæus.*—The Wheat-Grass is represented in Great Britain by five species; they have solitary spikelets, and with two nearly equal-sized glumes. The British examples are much more diminutive than the annual foreign species which are cultivated in this country for bread. The name is derived from the Latin *tritum*, and signifies thrashed or beaten, in allusion to the manner in which the corn is extracted from the ear.

FOUND on sandy sea-shores, where it is useful in binding the loose sand. It has no agricultural merits.

Common in England, Scotland, Ireland, France, Italy, Germany, Spain, Norway, Sweden, Portugal, West Asia, and North Africa.

Stem circular, upright, smooth, and having five or six lengthy smooth glaucous leaves, with smooth somewhat striated sheaths, the upper one shorter than its leaf; and having at its apex a brief membranous ligule. Inflorescence spiked, the spikelets oval in form, sessile, and in two alternate rows on a zigzag smooth rachis. Calyx consisting of two obtuse about equal-sized, smooth, six prominent-ribbed, glumes. Florets of two paleæ, the exterior one of basal floret smooth, five-ribbed, and of the same length as the calyx; inner palea having two green marginal ribs, and being minutely fringed. Length from fifteen to twenty-four inches. Root perennial and creeping.

Flowers at the commencement of July, and its seeds ripen about the middle of August.

The illustration is from a specimen gathered at Abergale, North Wales, by Mr. Joseph Sidebotham, of Manchester.

TRITICUM REPENS.

Smith. Hooker and Arnott. Parnell. Sinclair.
Deakin. Koch. Lindley. Greville. Babington Willdenow.
Knapp. Schrader. Host. Leers. Schreber Ehrhart. Hudson.
Withering. Hull. Relhan. Sibthorp. Abbot. Martyn.
Kunth. Macreight.

PLATE LXV.—A.

Triticum littorale,	Host.
" *junceum,*	Relhan.
Agropyrum repens,	Beauvais. Lindley.
Elymus arenarius,	Hudson.

The Creeping Wheat-Grass, or Couch-Grass.

Triticum—Wheat.   *Repens*—Creeping.

ONE of the most troublesome weeds that the farmer has to encounter, being difficult to eradicate from the soil having long creeping roots, which branch out in every direction.

It is common everywhere in England, Scotland, and Ireland. Also a native of Iceland, Russia, Norway, Sweden, Switzerland, Italy, France, Germany, Spain, Portugal, and the United States; yet not found above the altitude of six hundred feet.

Stem upright, circular, smooth, and striated, carrying five or six flat acute leaves with smooth striated sheaths; upper one shorter than its leaf, and having a very brief blunt ligule at its apex. Inflorescence spiked. Spike upright; spikelets oval, of four or five awnless florets, and placed alternately on the zigzag rachis in rows. Calyx consisting of two equal, acute glumes. Florets of two paleæ, exterior one of basal floret five-ribbed, harsh, and acute. Inner palea minutely dentated, and having two green marginal ribs. Length from twelve to twenty-four inches. Root perennial and creeping.

Flowers at the commencement of July, and ripens its seed in six weeks.

There is a common variety known as variety *aristatum*, which is frequently mistaken for *Triticum caninum*, but is distinguished in the glumes having five ribs.

The specimen illustrated was gathered at Beeston, near Nottingham.

TRITICUM CANINUM.

HUDSON. SMITH. HOOKER AND ARNOTT. PARNELL.
SINCLAIR. DEAKIN. KOCH. GREVILLE. LINDLEY. BABINGTON.
KNAPP. SCHRADER. HOST. OEDER. WITHERING. HULL. RELHAN.
ABBOT. KUNTH. MACKREIGHT.

PLATE LXV.—B.

Triticum biflorum,	MITTEN.
" *alpinum,*	DON.
Elymus caninus,	LINNÆUS. WILLDENOW.
" "	LEERS. EHRHART.
Agropyrum caninum,	BEAUVAIS. LINDLEY.

The Fibrous-rooted Wheat-Grass.

Triticum—Wheat.   *Caninum*—Dog.

THE Fibrous-rooted Wheat-Grass, or Bearded Wheat-Grass, is a valuable and early Grass.

Growing usually in damp shady places, yet thriving when cultivated in fields.

Common in England, Scotland, and Ireland.

Native of Siberia, Iceland, Norway, Sweden, Lapland, Germany, Switzerland, Italy, France, Spain, Portugal, and the United States.

Stem slender, upright, circular, and polished, having four or five broad, lanceolate acute, dark green, shining leaves, with smooth striated sheaths, upper one extending beyond its leaf, and having a very brief blunt ligule at its apex. Inflorescence spiked. Spike lengthy and delicate. Spikelets oval, sessile, placed in two rows on the zigzag rachis, and of four or five awned florets. The calyx composed of two rough, awned, three-ribbed, equal-sized glumes. Florets of two paleæ, exterior one of basal floret hirsute, five-ribbed, of same length as glume, and crowned with a lengthy slender awn. Inner palea membranous, and having two green marginal ribs. Length from twenty-four to forty-eight inches. Root perennial and fibrous.

Flowers at the commencement of July, and ripens its seed in a month.

The specimen illustrated was gathered at Congleton, by Mr. Joseph Sidebotham, of Manchester.

BRACHYPODIUM SYLVATICUM. B. PINNATUM.

BRACHYPODIUM SYLVATICUM.

BEAUVAIS. HOOKER AND ARNOTT. LINDLEY. DEAKIN. KOCH. BABINGTON.

PLATE LXVI.—A.

Festuca sylvatica,	SMITH. SINCLAIR. HUDSON. MARTYN.
" "	KNAPP. DICKSON. LIGHTFOOT.
" "	WITHERING. RELHAN. SIBTHORP.
" *gracilis,*	MŒNCH. SCHRADER.
Bromus sylvatica,	POLLICH. SMITH. HULL. HOOKER.
" "	SINCLAIR. POURRET. HOST.
" *gracilis,*	WEIGEL. ROTH. WILLDENOW.
" "	EHRHART.
Triticum sylvaticum,	MŒNCH. PARNELL. KUNTH.
" "	MACREIGHT.

The Slender False Brome-Grass.

Brachypodium—Short foot.   **Sylvaticum**—A wood.

BRACHYPODIUM. *Beauvais.*—The False Brome-Grass is named from the Greek, and signifies *short-footed.* This genus is intermediate between *Bromus* and *Triticum* There are two British examples.

OF no agricultural use, growing in damp shady situations; common in England, Scotland, and Ireland.

Native of France, Italy, Germany, Switzerland, and Russia.

Stem upright, circular, and smooth, bearing four or five broad sharp-pointed polished leaves, with hirsute striated sheaths, upper leaf extending beyond its sheath, and having a blunt hirsute ligule at its apex. Joints hairy, and four in number. Inflorescence racemed. Spikelets lengthy and cylindrical, generally of ten awned florets, placed alternately in two rows on the rachis. Calyx of two somewhat unequal, acute, hirsute, seven-ribbed glumes. Florets of two paleæ, exterior one of basal floret extending somewhat beyond the calyx; hirsute, seven-ribbed, with a long straight harsh awn. Length from twelve to twenty-four inches. Root perennial and fibrous.

Flowers at the commencement of July, and ripens its seed at the end of the same month.

The specimen from which the illustration is taken was gathered at Highfield House.

BRACHYPODIUM PINNATUM.

Beauvois. Hooker and Arnott. Lindley. Deakin. Babington.

PLATE LXVI.—B.

Festuca pinnata,	Hudson. Smith. Sinclair. Knapp.
" "	Dickson. Schrader. Relhan.
" "	Sibthorp. Abbot.
Bromus pinnatus,	Linnæus. Smith. Willdenow.
" "	Relhan. Sinclair. Pollich.
" "	Weigel. Host. Leers.
" "	Hudson. Hull.
Triticum pinnatum,	Mœnch. Parnell. Kunth.
" "	Macreight.

The Heath False Brome-Grass.

Brachypodium—Short foot.   *Pinnatum*—Feathered.

A USELESS agricultural Grass, growing on commons and on heathy situations, and preferring a chalk soil.

Frequent in the counties of Devon, Oxford, Cambridge, Nottingham, Bedford, Somerset, Dorset, Sussex, Kent, Suffolk, Norfolk, Gloucester, Worcester, Leicester, York, and Cumberland.

Native of Norway, Sweden, Italy, Germany, France, Spain, and Portugal.

Very subject to **variety**.

Stem delicate, upright, circular, and smooth, bearing four or five lengthy linear rough leaves, with striated sheaths; upper one shorter than its leaf. Ligules brief. Joints hirsute. Inflorescence racemed. Spikelets upright, long, and linear, generally of ten awned florets and two glumes. Glumes smooth, unequal, and seven-ribbed. Floret of two paleæ, exterior one of basal floret longer than the large glume.

Inner palea shorter. Awns shorter than their florets. Styles two. Stigmas plumose. Filaments three. Anthers notched at either extremity. Length thirty-six inches. Root perennial and creeping.

Flowers at the commencement of July.

Var. *gracile.*—More slender, and with shorter spikelets. Found in Kent.

Var. *cæspitosum.*—Spikelets small, leaves very narrow. Found near Bath.

Var. *compositum.*—Spikelets rising from the rachis in threes. Found in Yorkshire.

Var. *hispidum.*—Glume and florets hirsute. Found in Yorkshire.

Var. *hirsutum.*—Glume and florets hirsute, awns short, raceme erect. Found in Yorkshire.

The specimen illustrated was gathered on Mapperly Plains, near Nottingham.

LOLIUM PERENNE   L. MULTIFLORUM.

LOLIUM PERENNE.

LINNÆUS. HOOKER AND ARNOTT. SMITH. PARNELL.
DEAKIN. LINDLEY. SINCLAIR. GREVILLE. KOCH. BABINGTON.
WILLDENOW. RELHAN. KNAPP. MARTYN. GRAVES. SCHRADER. HOST.
SCHREBER. LEERS. OEDER. EHRHART. HUDSON. WITHERING.
SIBTHORP. KUNTH. MACREIGHT.

PLATE LXVII.—A.

Lolium tenue,   LINNÆUS. WILLDENOW.
" *rubrum,*   GERARDE.

The Perennial Rye-Grass.

Lolium—Darnel.   *Perenne*—Perennial.

LOLIUM. *Linnæus.*—This genus is known as Rye-Grass. There are three British examples.

A USEFUL agricultural Grass, and common throughout the whole of Britain. Native also of Lapland, Norway, Sweden, Switzerland, Russia, France, Italy, Spain, Portugal, Germany, United States, Northern Africa, and Western Asia.

Stem upright, circular, polished, and minutely striated, bearing six or seven lanceolate, acute, flat, deep green leaves, with smooth striated sheaths, the upper sheath extending beyond its leaf, and having a brief blunt ligule at its apex. Joints four or five. Inflorescence spiked. Spike upright and compressed. Spikelets sessile, and arranged alternately in two rows of six to twelve awnless florets. Calyx of one glume, smooth, five-ribbed, oblong-lanceolate, and shorter than the spikelets. Florets of two equal-sized paleæ, exterior one of basal floret five-ribbed, membranous, and shorter than the glume. Filaments slender. Styles brief. Stigmas plumose on the upper side. Anthers cloven at either extremity. Length from fifteen to twenty-four inches. Root perennial and fibrous.

Flowers in the middle of June, and ripens its seed at the commencement of July.

Var. *angustifolium*.—Tall and slender; leaves narrow and long.
Var. *racemosum*.—Having the spikelets pedunculated.
Var. *tenue*.—Dwarf, with spikelets of three or four florets.

The specimen illustrated was gathered at Highfield House.

LOLIUM MULTIFLORUM.

Lowe. Hooker and Arnott. Koch.

PLATE LXVII.—B.

Lolium italicum, BRAUN. BABINGTON.
" *perenne*, var. *italicum*, PARNELL.

The Bearded Rye-Grass, or Italian Rye-Grass.

Lolium—Darnel.   *Multiflorum*—Many-flowered.

SUPPOSED to have been introduced into England from Italy.
Stem upright, rough, and striated, bearing four or five lanceolate, flat, acute leaves, with harsh sheaths, upper one longer than its leaf. Inflorescence spiked. Spike from five to eight inches long. Spikelets from fourteen to twenty in number, composed of one glume, and from seven to eleven awned florets, the terminal one having two glumes. Glume linear-lanceolate. Florets of two equal-sized paleæ, five-ribbed. Styles two. Stigmas lengthy and plumose. Filaments three. Anthers lengthy, narrow, and notched at either extremity. Length from twenty-five to sixty inches. Root perennial and fibrous.

Flowers at the commencement of July.

A most valuable agricultural Grass, when cultivated on a rich deep soil.

Var. *submuticum.*—With large spikelets and short awns.

Var. *ramosum.*—Branched.

The specimen illustrated was gathered at Beeston, near Nottingham.

LOLIUM TEMULENTUM.   LEPTURUS INCURVATUS.

LOLIUM TEMULENTUM.

LINNÆUS. HOOKER AND ARNOTT. SMITH. PARNELL.
DEAKIN. LINDLEY. SINCLAIR. KOCH. BABINGTON. SCHRADER. HOST.
WILLDENOW. KNAPP. SCHREBER. LEERS.
OEDER. EHRHART. BULL. HUDSON. WITHERING. HULL. ABBOT.
RELHAN. KUNTH. MACREIGHT.

PLATE LXVIII.—A.

Lolium arvense,	WITHERING. HULL. SMITH. LINDLEY.	
" "	HOOKER. KNAPP. SCHRADER. HOST.	
" "	RELHAN. KUNTH. MACREIGHT.	
" *album*,	GERARDE. RAY.	
" *verum*,	GESNERI. MORISON.	

Darnel or Bearded Rye-Grass.

Lolium—Darnel.   *Temulentum*—Drunken, in allusion to the poisonous, sleepy property of the seeds.

NOT only a useless Grass, but a noxious weed, especially in corn-fields.

The seeds are said to be poisonous.

Not common in Scotland and Ireland, yet frequent in England, and found in the counties of Devon, Sussex, Kent, Essex, Cambridge, Suffolk, Bedford, Worcester, Nottingham, York, Durham, and Northumberland. In Wales in Carnarvonshire, and the Island of Anglesea. Also a native of France, Italy, Germany, Japan, South America, Norway, Sweden, and North Africa.

Stem circular, having four flat, lanceolate, acute, rough, minutely-toothed leaves, with smooth striated sheaths, the upper one being shorter than its leaf, and having a brief blunt ligule at its apex. Joints smooth, and four in number. Inflorescence spiked. Spike upright. Spikelets sessile, placed alternately in two rows of four or five awned florets. Rachis rough and zigzag. Calyx consisting of one glume, which is lengthy and narrow, and having eight ribs. Florets of two paleæ, the exterior one of basal floret seven-ribbed. Apex bifid, and having a white harsh awn, more than half the length of the palea, and commencing behind the bifid apex. Inner palea having two green marginal ribs. The glume longer than the spikelet, and smooth, except on the edges. Length twenty-four inches. Root annual and fibrous.

Dr. Parnell describes a variety which grows amongst and is equally common with the ordinary form, but which is stouter, and the awns of the florets longer. He has named it variety *longiaristatum*.

This species flowers at the commencement of July, and ripens its seeds in a month.

The illustration is from a specimen gathered near Hyde, in Cheshire, by Mr. Joseph Sidebotham, of Manchester.

LEPTURUS INCURVATUS.

TRIN. HOOKER AND ARNOTT. KOCH. MACREIGHT.
KUNTH. BABINGTON.

PLATE LXVIII.—B.

Rottbœllia incurvata,	LINNÆUS. SMITH. PARNELL.
" "	HOOKER. WILLDENOW. KNAPP.
" "	SCHRADER. HOST. OEDER.
" "	CAVANILLES. WITHERING. HULL.
" "	RELHAN. DEAKIN.
" *filiformis*,	ROTH. DON.
" *incurvatus*, var. *filiformis*,	HOOKER. PARNELL.
Lepturus filiformis,	TRIN. MACREIGHT. KUNTH.
Ophiurus incurvatus,	BEAUVOIS. LINDLEY.
Ægilops incurvata,	LINNÆUS. HUDSON.

The Curved Sea Hard-Grass.

Lepturus—Slender-spiked.   *Incurvatus*—Bowed down.

LEPTURUS. *Brown.*—The Hard-Grass derives its name from the Greek, and signifies *slender-tailed*, on account of the slender spikes. There is only a solitary British example.

A GRASS growing in salt marshes near the sea, and of no agricultural use.

In England it is found in the counties of Devon, Somerset, Sussex, Kent, Essex, Norfolk, Gloucester, Durham, and Northumberland. In Wales—in Denbigh, Flint, and the Island of Anglesea. In Scotland—along both the east and west coasts. In Ireland—common along the coasts.

Abroad it is met with along the shores of the Mediterranean.

Stem circular, polished, striated, base decumbent and bent at the joints, bearing six or seven narrow, acute, involute leaves, with smooth, striated, inflated sheaths, having a very brief blunt ligule at the apex. Inflorescence spiked. Spike lengthy and cylindrical. Spikelets alternate on the rachis. Calyx of two glumes, which are four-ribbed, lanceolate, acute, compact, and only spreading whilst in flower. Florets of two paleæ, somewhat shorter than the glumes, linear, membranous, and ribless. Filaments capillary. Anthers cloven at either extremity, and pendulous. Styles brief. Stigmas plumose. Length from three to six inches. Root annual and fibrous.

Flowers towards the close of July, and ripens its seed in the middle of August.

There is a slender erect-growing variety found near Aberlady, which is known as var. *filiformis*.

I am indebted to Mr. Joseph Sidebotham, of Manchester, and to Dr. Wilson, of Nottingham, for specimens.

The illustration is from a specimen gathered at Southsea, by Mr. T. Coward.

KNAPPIA AGROSTIDES.   SPARTINA STRICTA.
LXIX

KNAPPIA AGROSTIDEA.

SMITH. PARNELL. HOOKER AND ARNOTT. WITHERING. BABINGTON.

PLATE LXIX.—A.

Agrostis minima,	LINNÆUS. WILLDENOW. SMITH.
" "	HUDSON. STILLINGFLEET.
Chamagrostis minima,	SCHRADER. LINDLEY.
" "	BORCKHAUSEN. KOCH. MACREIGHT.
Sturmia minima,	HOPPE.
Mibora verna,	BEAUVOIS. REICHENBACH.
Gramen minimum,	DALECHAMPS. BAUHIN.

The Early Knappia.

Knappia—Named in honour of an English botanist, Mr. Knapp.
Agrostidea—........?

KNAPPIA.—A diminutive sea-side Grass, of which only one species is known, the *Knappia agrostidea*. Named after Mr. Knapp.

A DIMINUTIVE sea-side Grass, growing in sandy pastures. Of no agricultural use.

Anglesea and Jersey; a native also of France and central Europe.

Stem slender, upright, and having two or three narrow, blunt, smooth, channeled leaves, with smooth compressed sheaths, the upper sheath extending beyond its leaf. Inflorescence racemed, unilateral; rachis smooth. Spikelets briefly stalked, and of two glumes and one floret; glumes equal, obtuse, smooth, and destitute of lateral ribs. Floret of one palea, shorter than the glumes, blunt, hirsute, and white; apex ragged. Styles two, brief. Stigmas slender, lengthy, and plumose. Stamens three. Length from two to four inches. Root annual and fibrous.

Flowers in March and April, and ripens its seed in eight or nine weeks.

The specimen for illustration was gathered at Holyhead, by Mr. Joseph Sidebotham, of Manchester.

SPARTINA STRICTA.

SMITH. PARNELL. HOOKER AND ARNOTT. KUNTH. KOCH. LINDLEY. BABINGTON. DEAKIN. MACREIGHT.

PLATE LXIX.—B.

Dactylis stricta,   LINNÆUS. SMITH. KNAPP.
"   "   WITHERING. SOLANDER. WILLDENOW.
"   *cynosuroides,*   HUDSON. LOEFLING.

The Twin-spiked Cord-Grass.

Spartina—Named from its resemblance to *Lygeum spartum.*
Stricta—Close.

SPARTINA.—Erect-growing; spike compound. There are two British examples; name derived from the Greek.

AN interesting, although useless, erect-growing Grass; found on muddy salt marshes, on the east and south-east coasts of England.

A native of England, France, and Italy.

Stem striated, smooth, and sheathed to the apex, bearing numerous involute, rigid, pointed, smooth leaves, with smooth striated sheaths, the upper one extending beyond its leaf. Ligule brief, blunt, and ragged. Inflorescence of two or three spikes. Spikelets alternate, laterally compressed, and consisting of two glumes and one floret; glumes very unequal, hirsute, destitute of lateral ribs; inner glume largest. Floret of two unequal paleæ, exterior one shortest, hirsute, and destitute of lateral ribs. Inner palea having two delicate ribs. Stamens three; stigmas plumose; anthers upright and linear; base cloven; apex somewhat pointed; filaments lengthy and plumose.

Length from ten to twenty inches. Root perennial, fibrous, and creeping.

Flowers in August, and ripens its seed in the second week of September.

The specimen for illustration was gathered at Camsshore, near Fareham, by Mr. W. L. Notcutt.

SPARTINA ALTERNIFLORA.

SPARTINA ALTERNIFLORA.

LOISEL. SMITH. HOOKER AND ARNOTT. PARNELL. KUNTH. BABINGTON. MACREIGHT.

PLATE LXX.

Spartina glabra, MUHLEMB.
" *lævigata*, LINK.
Trachynotia alternifolia, DE CANDOLLE.

The Many-spiked Cord-Grass.

Spartina—Named from its resemblance to *Lygeum spartum*.
Alterniflora—Alternate-flowered.

A ROBUST reed-like Grass, growing on muddy banks of rivers. Cattle are particularly fond of it; also used for thatching.

Found on the banks of the Itchen and Southampton river, where it is very common, but not found elsewhere. A native also of North America.

Stem smooth, striated, and sheathed to the apex, bearing numerous, somewhat erect, alternate, strong, flat, (except on edges, where involute,) leaves, with smooth striated sheaths, the upper one extending beyond its leaf. Ligule brief, blunt, and jagged. Joints numerous. Leaves frequently a foot in length, the upper ones extending beyond the apex of the flower spikes. Inflorescence consisting of from four to thirteen spikes, which are close and compact. Spikelets alternate. Rachis ending in a wavy point. Spikelets of two glumes and one floret; glumes exceedingly unequal, the inner one largest, membranous and lanceolate; inner one five-ribbed. Floret consisting of two paleæ, which are shorter than the glumes, and acute; outer palea three-ribbed. Stamens three. Filaments shorter than the floret and capillary. Anthers linear and erect, cloven at the base, and pointed at the apex.

Stigmas plumose. Length from eighteen to twenty-four inches. Root perennial and creeping.

Flowers in August and September, and ripens its seed in October.

The specimen for illustration was gathered near Southport, by **Mr. Joseph Sidebotham**, of Manchester.

CYNODON DACTYLON.

PERSOON. R. BROWN. SMITH. PARNELL. HOOKER AND ARNOTT.
KOCH. KUNTH. BABINGTON.
LINDLEY. SINCLAIR. DEAKIN. MACREIGHT.

PLATE LXXI.

Panicum dactylon,	SMITH. KNAPP. LINNÆUS.
" "	WILLDENOW. DICKSON.
" "	HUDSON. WITHERING. HULL.
Digitaria stolonifera,	SCHRADER.
Agrostis linearis,	RETZIUS. WILLDENOW.

The Creeping Finger Grass, or Creeping Dog's Tooth Grass.

Cynodon—Dog's tooth.   *Dactylon*—........?

CYNODON.—Spike compound. Only one British example, the *Cynodon dactylon;* named from the Greek.

A PRETTY and singular Grass, common on the south-west coast of Cornwall, growing amongst the sand, but not found elsewhere. Of no agricultural use.

A native of Italy, Spain, Portugal, Greece, Turkey, the Mediterranean Islands, United States, West Indies, North Africa, and West Asia.

Stem smooth, base procumbent and then erect, bearing four or five flat, rigid, acute, hirsute leaves, with smooth striated sheaths, the upper one extending considerably beyond its leaf, destitute of a ligule, but furnished with a tuft of hairs. Inflorescence digitate, linear, and purplish. Spikelets laterally compressed, of two glumes and one floret; glumes almost equal, acute, destitute of lateral ribs; keel dentate on the upper half. Florets of two paleæ, destitute of lateral ribs, dorsal rib hirsute. Stamens three; pistils two; stigmas plumose. Length from three to six inches. Root perennial and creeping.

Flowers in July and August, and ripens its seed at the end of September.

The specimen for illustration was gathered at Penzance, by Mr. R. T. Millett, of Penzance.

DIGITARIA SANGUINALIS.
LXXII

DIGITARIA SANGUINALIS.

Scopoli.   Parnell.   Hooker and Arnott.   Smith.   Babington.
Lindley.   Deakin.   Sinclair.

PLATE LXXII.

Panicum sanguinale,	Smith. Linnæus. Koch.
" "	Kunth. Knapp.
" "	Curtis. Schreber. Hull.
" "	Hudson. Withering.
" "	Willdenow. Martyn.
" "	Ehrhart. Macreight.
Syntherisma vulgare,	Schrader. Schreber.
Ischæmon vulgare,	Lobel. Gerarde.

The Hairy Finger Grass.

Digitaria—From a finger.   *Sanguinalis*—Of blood.

DIGITARIA.—Spike compound. Two British examples. Named from the Latin.

A HANDSOME but useless agricultural Grass, supposed to have been introduced. Occasionally met with in England.

Native of France, Italy, Germany, Switzerland, America, North Africa and the West Indies.

Stem branched; base decumbent, then erect, striated and polished; having four brief, flat, somewhat broad, rough leaves with hirsute sheaths, the upper one extending considerably beyond its leaf. Joints three. Inflorescence digitate; branches lengthy, erect, and linear; from three to nine in number. Spikelets dorsally compressed, oblong-lanceolate, of two very unequal glumes and two florets; basal glume diminutive; upper one downy and three-ribbed; basal floret of one palea, flat and oblong-lanceolate, with five smooth ribs; margins pubescent; upper floret of two equal-sized paleæ. Filaments three; anthers short, violet-coloured, and cloven at either extremity. Styles two, slender. Stigmas brief, plumose, and purplish. Length from six to eighteen inches. Root annual and fibrous.

Flowers in August, and ripens its seed at the end of September.

The specimen for illustration was gathered near Bolton, by Mr. Joseph Sidebotham, of Manchester.

DIGITARIA HUMIFUSA.

Persoon. Smith. Parnell. Hooker and Arnott. Babington.
Lindley. Deakin.

PLATE LXXIII.

Syntherisma glabrum,	Schrader.
Panicum glabrum,	Koch. Gaudin.
" "	Macreight. Kunth.
" *humifusum*,	Kunth.
" *sanguinale*,	Pollich.
Digitaria filiformis,	Koel.

The Glabrous Finger Grass.

Digitaria—From a finger.   *Humifusa*—Spreading on the ground.

A RARE Grass, having no agricultural merits. Found on sandy soil in Yorkshire, Norfolk, Suffolk, Surrey, and Sussex.

A native also of France, Italy, Switzerland, Prussia, Holland, and Belgium.

Stem ascending, polished, striated, having four somewhat broad, brief, flat leaves, with smooth striated sheaths; upper one extending considerably beyond its leaf. Joints three. Inflorescence digitate, lengthy, linear, and from two to four-branched. Spikelets oval, dorsally compressed, of two glumes and one floret; glumes equal, hirsute, and five-ribbed; inner one deep purple, the others only purplish. Floret of same length as glumes—reddish purple, of two almost equal paleæ, striated and polished. Filaments three; anthers brief, violet-coloured, and cloven at either extremity. Styles two, slender. Stigmas brief, plumose, and purplish. Length from four to nine inches. Root annual and fibrous.

Flowers in July and August, and ripens its seed in September.

The specimen for illustration was gathered at Weybridge, Surrey, by Mr. Joseph Sidebotham, of Manchester.

PHRAGMITES COMMUNIS.
LXXIV

PHRAGMITES COMMUNIS.

TRIN. HOOKER AND ARNOTT. KOCH. BABINGTON. MACREIGHT. KUNTH. NEES.

PLATE LXXIV.

Arundo phragmites,		LINNÆUS. SMITH. PARNELL.
"	"	GREVILLE. LINDLEY.
"	"	WILLDENOW. KNAPP. HOOKER.
"	"	SCHRADER. LEERS. EHRHART.
"	"	HUDSON. WITHERING. RELHAN.
"	"	SIBTHORP. DEAKIN.
"	vallatoria,	RAY. GERARDE.
"	vulgaris,	BAUHIN. SCHEUCHZER.
"	palustris,	MATTHIOLUS. CAMERARIUS.

The Common Reed.

Phragmites—An enclosure. *Communis*—Common.

PHRAGMITES. *Trin.*—An abundant giant Grass; growing near water. Panicle large and noble. Name derived from the Greek on account of its use as a material for enclosure. Only one British example, namely, *Phragmites communis.*

A COMMON, handsome, giant Grass; of no agricultural use, yet useful for thatching, for the foundation of plaster floors, for arrows, and various other purposes.

Abundant throughout England, Scotland, and Ireland; growing in ditches, and on the margins of water.

A native also of France, Italy, Germany, Spain, Portugal, Russia, Norway, Sweden, Denmark, Lapland, New Holland, America, and North Africa.

Stem upright, circular, smooth, and strong, bearing fifteen and upwards of broad, lanceolate, many-ribbed, smooth leaves, with roughish striated sheaths, which extend beyond their leaves, and are destitute of ligules. Joints fifteen, smooth and polished. Inflorescence compound-panicled; panicle large, drooping on one side, pale brown in colour. Spikelets numerous, spreading, and of three awnless florets. Calyx of two unequal, acute, narrow glumes, with a rib on either side; upper one situated on a brief peduncle. Florets of two paleæ, exterior one of basal floret lanceolate, three-ribbed, and twice the length of the large glume; inner palea short. The peduncle of the second floret having lengthy, white, silky hairs spreading in every direction, and giving a beautiful silky appearance to the large panicle. Length from sixty to seventy inches. Root perennial and creeping.

Flowers in August, and ripens its seed in September.

The specimen for illustration was procured at Highfield House.

ADDENDA.

LEERSIA ORYZOIDES. Swartz.—A rare Grass, growing in ditches and damp places. Henfield and Arundel, Essex; Mole River, Surrey; Boldre River, near Brockenhurst Bridge, Hants.

AGROSTIS INTERRUPTA. Linnæus.—Rare. Sandy pastures. Thetford. Closely allied to *Agrostis spica-venti*.

POA BORRERI. Hooker and Arnott.—*Glyceria conferta*, Fries; *Schlerochloa borreri*, Babington. South of Europe, in brackish places.

POA LAXA. Hœnck.—Rare. Ben Nevis, Loch na Gar, and Clova Mountains.

BROMUS RACEMOSUS. Linnæus.—Sandy situations. South of England. Scarcely different to *B. commutatus*.

BROMUS PATULUS. Koch.—Found by Mr. Gibson, near Hebden Bridge, Yorkshire. Closely allied to *B. arvensis*.

BROMUS SQUARROSUS. Linnæus.—Corn-fields. Surrey, Kent, Essex, and Somersetshire. An introduced species.

AVENA PLANICULMIS. Schrader.—Isle of Arran. Introduced.

ELYMUS ARENARIUS. Linnæus.—Sea-shores.

ELYMUS GENICULATUS. Curtis.—Very rare. Gravesend.

TRITICUM CRISTATUM. Schreber.—Rocks at the sea-side between Arbroath and Montrose. Rare.

TRITICUM LAXUM. Fries.—Sandy sea-shores. A doubtful species.

CONCLUSION.

IN the present Work there are several species not enumerated, some being of doubtful origin, and others so exceedingly rare as scarcely to be procured: they are mentioned briefly in the addenda. At the conclusion of this Work the author intended to have commenced a Natural History of those Foreign Grasses that were remarkable for their beauty, singularity, or economic values; this latter Work, however, will be deferred, as the author cannot devote the time requisite for this undertaking at the present moment. In order to distinguish the species of British Grasses, the author has appended a

COMPARATIVE ANALYSIS,

ARRANGED BY MR. RALFS.

ORDER I.—MONOGYNIA.   One style.

Stigma one	*Nardus.*
Stigmas two	Some Grasses.

CLASS II.—DIANDRIA.   Two styles.

ORDER II.—DIGYNIA.   Two styles.

Calyx single-flowered	*Anthoxanthum.*
Calyx two or more flowered	Some Grasses.

DIANDRIA.   *Digynia.*

ANTHOXANTHUM.

Calyx valves very unequal	*Odoratum.*

CLASS III.—TRIANDRIA.   Three stamens.

1. Flowers spiked	2
Flowers panicled	12
2. Flowers in unilateral spikes	3
Flowrs not unilateral	6

3. Calyx many-flowered . *Triticum.*
   Calyx one or two-flowered . 4
4. Styles united half way up *Spartina.*
   Styles distinct . 5
5. Glumes nearly equal . *Cynodon.*
   Glumes very unequal *Digitaria.*
6. Spikelets imbedded in the rachis *Rottbolia.*
   Spikelets not imbedded . 7
7. Spikelets two or more from same point 11
   Spikelets solitary . 8
8. Spikelets one-flowered . *Knappia.*
   Spikelets more than two-flowered . 9
9. Glume solitary, inclosing the spikelet between it and rachis . *Lolium.*
   Glumes two, their edges towards the rachis . 10
10. Florets equal . *Brachypodium.*
    Florets smaller upwards *Triticum.*
11. Spikelets one-flowered . *Hordeum.*
    Spikelets two or more flowered *Elymus.*
12. Calyx one-flowered . 13
    Calyx two or more flowered 23
13. Glumes with feathery awns . *Lagurus.*
    Glumes awnless, or awns not feathery . 14
14. Panicle dense, (subspiked,) florets mostly imbricated 15
    Panicle loose, florets not imbricated . 20
15. Corolla with tuft of hair at base . *Amophilla.*
    Corolla without hairs at base 16
16. Corolla with one or two valves of imperfect florets at base . *Phalaris.*
    Corolla without valves of imperfect florets at base . 17
17. Corolla awnless *Phleum.*
    Corolla awned . 18
18. Corolla of one valve, the awn basal . *Alopecurus.*
    Corolla of two valves, the awn terminal or dorsal 19
19. Glumes awned, awn of corolla terminal . *Polypogon.*
    Glumes awnless, awns of corolla dorsal . *Gastridium.*
20. Fruit inverted with the hardened corolla; corolla awnless 21
    Fruit not inverted with the corolla; corolla often awned 22
21. Corolla with a small valve at the base *Phalaris.*
    Corolla without a valve at the base . *Milium.*
22. Corolla with long hairs at base *Calamagrostis.*
    Corolla without hairs at base *Agrostis.*
23. Florets with a pinnated bractea *Cynosurus.*
    Florets without a pinnated bractea . 24
24. Calyx with not more than two perfect florets . 25
    Calyx with more than two perfect florets 35
25. Panicle dense, in a simple or compound spike 26

	Panicle lax, not spiked	29
26.	Some of the florets with stamens only	27
	Florets perfect	28
27.	Spike compound, florets without bristles	*Panicum.*
	Spike simple, florets with a bristly involucre	*Setaria.*
28.	Calyx valves nearly equal, styles united	*Sesleria.*
	Calyx valves unequal, styles distinct	*Airochloa.*
29.	Corolla awnless	30
	Corolla awned	31
30.	Glumes truncate	*Catabrosa.*
	Glumes acute	*Melica.*
31.	Florets perfect	32
	Some of the florets with stamens only	33
32.	Corolla invested with seed	*Avena.*
	Corolla not invested with seed	*Aira.*
33.	Florets three, perfect one with two stamens	*Hierochlæ.*
	Florets two, perfect one with three stamens	34
34.	Upper floret barren, lower one perfect	*Holcus.*
	Lower floret barren, upper one perfect	*Arrhenatherum.*
35.	Panicle imbricated and simple	*Sesleria.*
	Panicle loose, or, if dense, branched	36
36.	Corolla with long hairs at base	*Arundo.*
	Corolla without hairs at base, or nearly so	37
37.	Spikelets in dense clusters at end of the branches	*Dactylis.*
	Spikelets not in dense clusters	38
38.	External valve of corolla with three nearly equal teeth	*Triodia.*
	External valve of corolla not three-toothed	39
39.	Corolla awned or pointed	40
	Corolla obtuse, not awned	42
40.	Corolla with a twisted dorsal awn, upper florets mostly imperfect	*Avena.*
	Corolla pointed, or with an awn terminal, or nearly so, florets perfect	41
41.	Corolla with a terminal awn, or pointed	*Festuca.*
	Corolla with an awn just below the bifid extremity	*Bromus.*
42.	Spikelets cordate, (pendulous,) seed coated by the corolla	*Briza.*
	Spikelets not cordate, seeds free	43
43.	Spikelets linear or sub-cylindrical, not webbed	*Glyceria.*
	Spikelets ovate or oblong, often webbed	*Poa.*

TRIANDRIA. *Monogynia.*

NARDUS.

Florets spiked, unilateral . . . . *Stricta.*

TRIANDRIA. *Digynia.*

ALOPECURUS.

1. Glumes not united . . . . *Bulbosus.*
   Glumes united at base 2
2. Spike ovate, inflated sheath of upper leaf thrice as long as
   the leaf . . . . *Alpinus.*
   Spike cylindrical, sheath not thrice as long as upper leaf . 3
3. Culm erect, glumes acute 4
   Culm ascending, glumes obtuse 5
4. Spike obtuse, awn twice the length of corolla . . . *Pratensis.*
   Spike acute, awn more than twice length of corolla . *Agrestis.*
5. Awn as long as the glumes . . . . *Fulvus.*
   Awn longer than the glumes . . . . *Geniculatus.*

PHALARIS.

Spike dense, ovate . . . . *Canariensis.*
Panicle branched . . . . *Arundinacea.*

AMMOPHILA.

Glumes acute . . . . *Arundinacea.*

PHLEUM.

1. Glumes awnless, twice as long as the corolla . . *Arenarium.*
   Glumes mostly awned, not twice as long as the corolla . 2
2. Glumes naked or downy (not ciliated) at the back . 3
   Glumes ciliated at the back 4
3. Culm mostly branched, glumes wedge-shaped . . . *Asperum.*
   Culm simple, glumes lanceolate . . . . *Bœhmeri.*
4. Glumes lanceolate, gradually tapering . . . . . *Michelii.*
   Glumes truncated 5
5. Spike ovate-oblong, awn as long as the glume . . *Alpinum.*
   Spike cylindrical, awn shorter than the glume . *Pratense.*

LAGURUS.

Awns long . . . . *Ovatus.*

MILIUM.

Florets glabrous . . . . *Effusum.*

GASTRIDIUM.

Awns twice as long as the glumes . . . . *Lendigerum.*

POLYPOGON.

Awns as long as the calyx, root creeping . . . . *Littoralis.*
Awn much longer than the calyx, root fibrous . *Monspeliensis.*

CALAMAGROSTIS.

1. Corolla with a terminal awn, panicle loose . . *Lanceolata.*
   Corolla with a dorsal or basal awn, panicle close . 2

2. Flowers without a rudiment of a second floret . . *Epigejos.*
Flower with a minute pedicel, bearing a tuft of hair (a rudiment of a second floret) . . . 3
3. Hairs as long as the corolla, awn near the base . *Lapponica.*
Hairs shorter than the corolla, awn above the middle . *Stricta.*

AGROSTIS.

1. Inner valve of corolla wanting, or minute   2
   Inner valve of corolla not minute .   3
2. Leaves linear, awn dorsal . *Canina.*
   Leaves setaceous, awn basal . . . *Setacea.*
3. Awn very long and terminal, florets with a barren pedicel at base . . . . . *Spica-venti.*
   Awn none, or short and dorsal, no barren pedicel . 4
4. Ligule short and truncate, outer valve of corolla three-nerved . . . . . *Vulgaris.*
   Ligule oblong, outer valve of corolla five-nerved . *Alba.*

CATABROSA.

Branches whorled . . *Aquatica.*

AIROCHLOA.

Glumes shorter than the florets . . *Cristata.*

AIRA.

1. Awn clavate . *Canescens.*
   Awn not clavate . . 2
2. Leaves linear, awns not or but little longer than corolla   3
   Leaves setaceous, awns longer than corolla . . 4
3. Awn basal, branches rough . *Cæspitosa.*
   Awn dorsal, branches smooth . *Alpina.*
4. Florets hairy at base *Flexuosa.*
   Florets scarcely hairy at base . 5
5. Panicle close, awn basal . *Præcox.*
   Panicle spreading, awn dorsal . *Caryophyllea.*

MELICA.

1. Panicle drooping, florets not longer than the calyx   2
   Panicle erect, florets much longer than the calyx   3
2. Spikelets with two perfect florets . *Nutans.*
   Spikelets with one perfect floret . *Uniflora.*
3. Leaves much shorter than the purplish panicle . *Cærulea.*
   Leaves much longer than the pale panicle *Depauperata.*
   (M. cærulea β. *Hooker.*)

HOLCUS.

Awn longer than the calyx, joints of culm downy . *Mollis.*
Awn not longer than the calyx, joints not downy *Lanatus.*

ARRHENATHERUM.

Spikelets greenish brown — *Avenaceum.*

HIEROCHLOE.

Florets awnless — *Borealis.*

SESLERIA.

Spike bluish, ovate — *Cærulea.*

PANICUM.

Spikes alternate — *Crus-galli.*

SETARIA.

Bristles of involucre with reflexed teeth — *Verticillata.*
Bristles of involucre with erect teeth — *Viridis.*

GLYCERIA, (POA, *Hooker.*)

1. Florets with seven or more ribs — 2
   Florets with not more than five ribs — 3
2. Panicle much branched, plant four feet high — *Aquatica.*
   Panicle slightly branched, plant not more than three feet high — *Fluitans.*
3. Panicle compact, rigid — 4
   Panicle spreading, not rigid — 6
4. Root creeping — *Maritima.*
   Root fibrous — 5
5. Florets four or five, five-ribbed, plant procumbent — *Procumbens.*
   Florets six or seven, nearly ribless, plant erect, or ascending — *Rigida.*
6. Panicle reflexed in fruit — *Distans.*
   Panicle not reflexed in fruit — *Borreri.*

POA.

1. Florets webbed — 2
   Florets not webbed — 7
2. Stems bulbous at base — *Bulbosa.*
   Stems not bulbous at base — 3
3. Culm much compressed, spikelets three or more flowered — *Compressa.*
   Culm round, or but little compressed, spikelets three or four-flowered — 4
4. Florets obscurely ribbed, panicle somewhat drooping — 5
   Florets five-ribbed, panicle not drooping — 6
5. Ligules lanceolate — *Laxa.*
   Ligules short and truncate — *Nemoralis.*
6. Culm and sheath smooth, root creeping — *Pratensis.*
   Culm and sheath rough, root fibrous — *Trivialis.*

7. Ligules short and truncate, panicle somewhat drooping   *Nemoralis.*
   Upper ligule oblong, acute, panicle not drooping   8
8. Stem spreading, procumbent at the base   *Annua.*
   Stem nearly erect   *Alpina.*

TRIODIA.

Ligule, a tuft of hairs   *Decumbens.*

BRIZA.

Glumes longer than the florets, ligule lanceolate   *Minor.*
Glumes shorter than the florets, ligule very short   *Media.*

DACTYLIS.

Panicle secund   *Glomerata.*

CYNOSURUS.

Spike ovate   *Echinatus.*
Spike linear   *Cristatus.*

FESTUCA.

1. Glume one   *Uniglumis.*
   Glumes two   2
2. Leaves auricled   *Bromus giganteus.*
   Leaves not auricled   3
3. Florets monandrous, shorter than their awns   4
   Florets triandrous, awnless, or as long as their awns   5
4. Culm leafy in its upper part   *Myurus.*
   Culm leafless above   *Bromoides.*
5. Raceme spiked   *Loliacea.*
   Panicle branched   6
6. Lower leaves setaceous or involute, pedicels naked   7
   Leaves linear, pedicels tufts of hair at end   10
7. Culm square, not a foot high   8
   Culm round   9
8. Florets awned or pointed, edge of inner valve glabrous   *Ovina.*
   Florets awnless, edge of inner valve downy   *Vivipara.*
9. Root fibrous   *Duriuscula.*
   Root creeping   *Rubra.*
10. Spikelets not more than five-flowered   *Calamaria.*
   Spikelets more than five-flowered   11
11. Panicle much branched, root creeping   *Diandrus.*
   Panicle not much branched, root fibrous   *Pratensis.*

BROMUS.

1. Stamens two   *Diandrus.*
   Stamens three   2

2. Leaves auricled, glabrous . . . *Giganteus.*
   Leaves not auricled, mostly pubescent 3
3. Awn much longer than florets . 4
   Awn not much longer than florets . . 5
4. Panicle erect in flower, awn twice as long as the florets *Maximus.*
   Panicle drooping, awn not twice as long as the florets *Sterilis.*
5. Florets pubescent 6
   Florets glabrous 8
6. Florets about eight, remote, and longer than the awn *Asper.*
   Florets nine or more, crowded, as long as the awn 7
7. Leaves slightly hairy, panicle spreading . *Velutinus.*
   Leaves very pubescent, panicle erect, close *Mollis.*
8. Panicle drooping in fruit . . 9
   Panicle erect 11
9. Spikelets lanceolate, awns straight *Arvensis.*
   Spikelets ovate, awns not straight . 10
10. Awns remarkably spreading, leaves pubescent *Squarrosum.*
    Awns not spreading, leaves slightly hairy *Secalinus.*
11. Root leaves much narrower than the cauline . *Erectus.*
    Root leaves not narrower than the cauline *Racemosus.*

AVENA.

1. Spikelets drooping, florets not larger than the calyx . 2
   Spikelets erect, florets mostly longer than the calyx 3
2. Florets terminated by two bristles . *Strigosa.*
   Florets not terminated by two bristles . *Fatua.*
3. Glumes very unequal . . *Flavescens.*
   Glumes nearly equal . . 4
4. Leaves downy, spikelets two or three-flowered . *Pubescens.*
   Leaves not downy, spikelets more than three-flowered 5
5. Lower leaves involute . . . *Pratensis.*
   Leaves flat . . . 6
6. Sheaths flat, lower part of culm two-edged . *Planiculmis.*
   Sheaths round, culm round . . . *Alpina.*

ARUNDO.

Florets purplish, culm five feet high . . *Phragmites.*

ELYMUS.

1 Leaves flat, florets awned . *Europæus.*
  Leaves involute, florets awnless . . 2
2. Spike drooping, glumes longer than the florets . *Geniculatus.*
   Spike erect, glumes not longer than the florets . *Arenarius.*

HORDEUM.

1. Glumes all setaceous . *Pratense.*

Glumes not *all* setaceous . . . 2
2. Glumes of lateral florets setaceous, of central floret lanceolate *Murinum*.
Inner glumes of lateral floret not setaceous, the rest
setaceous . . . . *Maritimum*.

TRITICUM.

1. Spikelets unilateral . *Loliaceum*.
Spikelets distichous . . . 2
2. Glumes and florets obtuse, leaves involute . . *Junceum*.
Glumes and florets awned or pointed, leaves flat . . 3
3. Glumes scarcely ribbed, spikelets crowded . *Cristatum*.
Glumes ribbed, spikelets not crowded . . 4
4. Root fibrous . . . *Caninum*.
Root creeping . . . *Repens*.

BRACHYPODIUM.

Spike drooping, awns longer than the florets . *Sylvaticum*.
Spike erect, awns shorter than the florets . . *Pinnatum*.

LOLIUM.

1. Spikelets much longer than the calyx . . *Perenne*.
Spikelets not longer than the calyx . . . 2
2. Florets with long rigid awns . . . *Temulentum*.
Florets with short soft awns . . *Arvense*.

ROTTBŒLLIA.

Glumes united . . . . *Incurvata*.

KNAPPIA.

Root fibrous, florets very hairy . . *Agrostidea*.

SPARTINA.

Florets very hairy . . *Stricta*.
Florets glabrous . *Alterniflora*.

CYNODON.

Florets glabrous, longer than the glumes . *Dactylon*.

DIGITARIA.

Leaves pubescent *Sanguinalis*.
Leaves glabrous *Humifusa*.

Dr. Parnell gives the following analysis:—

ALOPECURUS.—Stem rough *Agrestis*.
Stem smooth . 1

1. Upper leaf much shorter than its sheath   2
   Upper leaf about equal in length to its sheath   3
2. Awn projecting more than half its length beyond floret   Alpinus.
   Awn projecting not more than a third beyond floret   Pratensis.
3. Awn projecting half its length beyond floret   Geniculatus.
   Awn not projecting beyond floret   Fulvus.

PHLEUM.—Glumes awned   1
         Glumes acute, not awned   2
1. Awn not half the length of glume   Pratense.
   Awn more than half the length of glume   Alpinum.
2. Floret not half length of calyx   Arenarium.
   Floret more than half length of calyx   Michelii.

PHALARIS.—Base of floret with two membranous valves   Canariensis.
          Base of floret with two hairy valves   Arundinacea.

HORDEUM.—Glumes of middle spikelet fringed   Murinum.
         Glumes of middle spikelet not fringed   1
1. Inner glume of lateral spikelet very much dilated on one side Maritimum.
   Glumes not dilated   Pratense.

AGROSTIS.—Ligule of upper sheath very short   Vulgaris.
          Ligule of upper sheath long   1
1. Floret of two paleæ, sheaths roughish   Alba.
   Floret of one palea, sheaths smooth   Canina.

CALAMAGROSTIS.—Hairs shorter than floret   Stricta.
               Hairs longer than floret   Epigejos.

MELICA.—Calyx with one floret and rudiment of second   Uniflora.
        Calyx with two florets and rudiment of third   Nutans.

MOLINIA.—Outer palea five-ribbed   Depauperata.
         Outer palea three-ribbed   Cœrulea.

HOLCUS.—Awn of floret smooth   Lanatus.
        Awn of floret rough   Mollis.

AIRA.—Awns not protruding beyond the florets   1
      Awns protruding considerably beyond the florets   2
1. Awn arising from little above base of palea   Cæspitosa.
   Awn arising from little above centre of outer palea   Alpina.
2. Sheath of leaf rough from above downwards   Flexuosa.
   Sheath of leaf rough from below upwards   3
3. Panicle spreading   Caryophillea.
   Panicle close   Præcox.

AVENA.—Florets with two long bristles at summit   Strigosa.
       Florets without bristles at summit   Fatua.

COMPARATIVE ANALYSIS.

CYNOSURUS.—Outer palea terminating in a short awn, not half
    length of palea . . . . *Cristatus.*
    Outer palea terminating in a long awn as long as
    the palea . *Echinatus.*

POA.—Florets webbed . 1
    Florets not webbed . . 4
1. Upper leaf much longer than sheath . 2
   Upper leaf about as long or longer than sheath . 3
2. Ligule of upper sheath short and rounded *Pratensis.*
   Ligule of upper sheath long and pointed . *Trivialis.*
3. Ligule scarcely perceptible, outer palea five-ribbed *Nemoralis.*
   Ligule prominent, outer palea three-ribbed . *Compressa.*
4. Florets hairy at base . . 5
   Florets not hairy . 12
5. Outer palea three-ribbed . 6
   Outer palea five-ribbed . . 7
6. Panicle erect, upper leaf linear, folded . *Alpina.*
   Panicle drooping, upper leaf lanceolate, flat . *Laxa.*
7. Upper joint situated above centre of stem . 8
   Upper joint situated below centre of stem . 9
8. Second sheath not reaching to first joint *Polynoda.*
   Second sheath extending beyond first joint *Montana.*
9. Small glume reaching beyond base of third floret 10
   Small glume not reaching beyond base of second floret 11
10. Rachis and branches rough *Cæsia.*
    Rachis and branches smooth . *Annua.*
11. Rachis and branches rough to touch *Distans.*
    Rachis and branches smooth to touch . *Maritima.*
12. Glumes with a prominent lateral rib on each side *Procumbens.*
    Glumes without lateral ribs . . 13
13. Lower half of central rib of outer palea smooth 14
    Central rib of outer palea rough the whole length . 15
14. Summit of upper glume reaching to base of third floret *Rigida.*
    Summit of upper glume reaching to base of fourth floret *Loliacea.*
15. Outer palea three-ribbed . . *Sylvatica.*
    Outer palea seven-ribbed . . 16
16. Panicle compound, spikelets not a quarter of an inch in
    length . . . . . *Aquatica.*
    Panicle simple, spikelets an inch in length . *Fluitans.*

BUCKTUM.—Inflorescence racemed, approaching to a spike *Loliaceum.*
    Inflorescence panicled . . 1
1. Panicle simple . *Pratense.*
   Panicle compound . . 2
2. Awn considerably shorter than palea *Elatius.*
   Awn considerably longer than palea . *Giganteum.*

BROMUS.—Large glume seven-ribbed . . 1
   Large glume three-ribbed . . . 4
 1. Summit of upper glume midway between its base and
   summit of third floret 2
   Summit of upper glume midway between its base and
   summit of second floret . . . 3
 2. Florets and glumes hairy . . *Mollis.*
   Florets and glumes not hairy . . . *Racemosus.*
 3. Twice width of outer palea considerably more than length
   of palea . . . . *Secalinus.*
   Twice width of outer palea equal to length of palea . *Arvensis.*
 4. Awns of florets much longer than calyx . 5
   Awns of florets much shorter than calyx . . 6
 5 Spikelets drooping, awns longer than the florets . *Sterilis.*
   Spikelets erect, awns equal in length to the florets . *Diandrus.*
 6 Lower floret about one-third longer than the small glume . *Erectus.*
   Lower floret about twice the length of the small glume . *Asper.*

TRISETUM.—Radical leaves hairy . . 1
    Radical leaves not hairy . . *Pratense.*
 1. Ligule long and acute . *Pubescens.*
   Ligule very short and obtuse . *Flavescens.*

FESTUCA.—Awns much longer than the florets *Bromoides.*
    Awns much shorter than the florets . . 1
 1. Root fibrous, stem under the panicle rough *Ovina.*
   Root creeping, stem under the panicle smooth *Duriuscula.*

TRITICUM.—Spikelets long, on short footstalks . *Sylvaticum.*
    Spikelets short, without footstalks . 1
 1. Stem rough . . . *Cristatum.*
   Stem smooth 2
 2. Awns rather longer than the florets *Caninum.*
   Awns very short, or wanting . 3
 3. Rachis rough . . *Repens.*
   Rachis smooth . *Junceum.*

LOLIUM.—Florets awned, glume longer than the spikelet *Temulentum.*
   Florets not awned, glume shorter than the spikelet *Perenne.*

INDEX.

	PAGE
Acorus calamus	91
Ægilops incurvata	209
Agropyrum caninum	197
junceum	193
repens	195
Agrostis alba. *Linnæus*	58, 59
var. palustris	60
var. stolonifera	60
alpina	57
australis	39
canina. *Linnæus*	55, 58, 61
var.	57
capillaris	59, 61
compressa	59, 60
fascicularis	55
hispida	61
interrupta	225
linearis	217
littoralis	47
minima	221
mutabilis	57, 59
palustris	59
panicea	45
polymorpha	59, 61
pumila	61
rubra	39
setacea. *Curtis*	57
stricta	55
spica-venti. *Linnæus*	63
stolonifera	59, 61
stolonifera-latifolia	59

	PAGE
Agrostis sylvatica	59, 60
tenuifolia	55
tenuis	61
triaristata	45
ventricosa	39
vinealis	55
vulgaris. *Withering*	58, 61
var. cristata	62
var. pumila	62
Aira alpina. *Linnæus*	68, 69
var. vivipara	70
aquatica	65
var.	105
canescens. *Linnæus*	75
caryophyllea. *Linnæus*	71, 73, 77
cœrulea	79
cæspitosa. *Linnæus*	67, 68
var. brevifolia	68
var. longiaristata	68
var. vivipara	68
cristata	93
flexuosa. *Linnæus*	68, 69, 73
var. montana	74
lævigata	69
montana	73
præcox. *Linnæus*	77
scabro-setacea	73
setacea	73
Airochloa cristata	93
Alopecuros genuina	37

INDEX.

	PAGE
Alopecuros spica-rotundiore	37
Alopecurus agrestis. *Linnæus*	8, 11, 13, 14
alpinus. *Smith*	8, 9, 18, 28
bulbosus. *Linnæus*	13, 25
fulvus. *Smith*	8, 12, 14, 15, 18
geniculatus. *Linnæus*	8, 12, 14, 15, 17, 18
var.	15
maxima anglica	45
myosuroides	11
ovatus	9
paniceus	17, 45
pratensis. *Linnæus*	7, 12, 14, 18
ventricosus	39
Ammophila evenaria	23
arundinacea. *Host*	23
Anemagrostis spica-venti	63
Anthoxanthum odoratum. *Linnæus*	3, 92
Arrhenatherum avenaceum. *Beauvois*	89
var.	89
bulbosum	89
elatior	89
pallens	90
Arundo arenaria	23
aristatus	45
calamagrostis	49, 51
colorata	21
epigejos	49, 51
neglecta	53
monspeliensis	45
palustris	223
phragmitis	223
stricta	53
vallatoria	223
vulgaris	223
Avena alpina	177
bromoides	177
caryophyllea	71
elatior	89
fatua. *Linnæus*	175, 181
flavescens. *Linnæus*	175, 183

	PAGE
Avena nodosa	89
pratensis. *Linnæus*	175, 177
var. latifolium	178
var. longifolium	178
planiculmis	175, 177, 225
precatoria	89
pubescens. *Linnæus*	175, 179
sativa	181, 182
sesquitertia	179
strigosa. *Schreber*	175, 181, 182
Brachypodium pinnatum. *Beauvois*	201
var. cæspitosum	202
var. compositum	202
var. gracile	202
var. hirsutum	202
var. hispidum	202
var. sylvaticum. *Beauvois*	199
Briza aspera	137
maxima	37
media. *Linnæus*	135
Bromus agrestis	157
arvensis. *Koch*	167
asper. *Linnæus*	159
ciliatus	171
commutatus. *Schrader*	165
diandrus. *Curtis*	171
erectus. *Hudson*	157
var. hirsutus	158
giganteus	147
glomeratus	138
gracilis	199
grandiflorus	161
gynandrus	171
hirsutus	159
hordeaceus	169
littoreus	145
madritensis	171
maximus. *Desfontaines*	173
montanus	159
mollis. *Linnæus*	169
muralis	171
multiflorus	163

INDEX.

Bromus nemoralis	. . .	159
nemorosus	. . .	159
nanus	. . .	169
patulus	. .	225
perennis	. .	157
pinnatus	. .	120
polymorphus		163, 169
racemosus	. .	225
ramosus		159
rigidus	. . .	171
secalinus. *Smith*	.	163
var. velutinus	. .	164
var. vulgaris	. .	164
squarrosus	. .	225
sylvaticus	. .	199
sterilis. *Linnæus*		161, 173
triflorus	. . .	147
velutinus	. . .	163
vitiosus	. .	163
Bucetum elatius	. .	145
giganteum	. .	147
loliaceum	. .	143
pratense	. .	143
Calamagrostis arenaria	. .	23
eligegos. *Roth*	49, 51,	52, 53
lanceolata. *Roth*	.	49, 51
lapponica	. .	52
stricta. *Nuttall*	49,	50, 53
Catabrosa aquatica. *Beauvois*		65
var. littoralis	. .	66
Catapodium unilaterale	. .	117
Chamagrostis minima	. .	211
Chilochloa boehmeri	.	33
Corynephorus canescens	. .	75
Cynodon dactylon. *Persoon*	.	217
Cynosurus cœruleus	. .	95
cristatus. *Linnæus*	.	139
echinatus. *Linnæus*	139,	141
paniceus	. .	45
Dactylis cynosuroides	. .	213
glomerata. *Linnæus*	133,	134
stricta	. .	213
Danthonia decumbens	. .	131
Deschampsia cæspitosa	. .	67
Digitaria filiformis	. .	221

Digitaria humifusa. *Persoon*	.	221
sanguinalis. *Scopoli*		219
stolonifera	. .	217
Echinochloa crus-galli	.	97
Elymus arenarius	. .	226
caninus		197
europeus	. .	185
geniculatus	.	226
Fustuca arundinacea	. .	145
bromoides. *Linnæus*	.	153
var. nana	.	154
var. pseudo-myurus	.	154
calamaria	. .	151
cæsia	.	155
decumbens	.	133
distans	. .	105
dumetorum	. .	155
elatior. *Linnæus*	143,	145
var. variegatum	.	146
fluitans	.	107
var.	. .	143
gigantea. *Villars*	.	147
gracilis	. .	199
heterophylla	.	155
hirsuta	. .	155
loliacea	. .	143
madritensis	. .	171
myurus	. .	153
nemorum	. .	155
ovina. *Linnæus*	.	155
var. angustifolia	.	156
var. arenaria	.	156
var. cæsia	.	156
var. duriuscula	.	156
var. filiformis		156
var. humilis		156
var. hirsuta	.	156
var. rubra	.	156
var. vivipara	.	154, 155
pratensis. *Hudson*	108,	143
pseudo-myurus	.	153
pinnata	. .	201
rigida	. .	113
rubra	. .	155
sciuroides	. .	153

INDEX.

	PAGE
Festuca sylvatica. *Villars*	151, 199
thalassia	109
triflora	147
tenuifolia	155
uniglumis. *Solander*	149
vivipara	155
Gastridium australe	39
lendigerum. *Beauvois*	39, 40
Gramen alopecuroides-majus	7
arundinaceum	63
asperum	131
bulbosum-nodosum	89
caninum-nodosum	89
cristatum	139
fluviatile	107
geniculatum	99
junceum	75
miliaceum	41
miliaceum-vulgare	41
minima	211
segetale	67
secalinum	187
typhinum-minus	25
typhoides-minus	25
Glyceria aquatica	103
distans	105
fluitans	107
maritima	109
pedicellata	107
plicata	107
procumbens	111
rigida	113
Hierochloë borealis. *Rœmer*	91
odorata	91
Holcus avenaceus	89
borealis	91
lanatus. *Linnæus*	89
mollis. *Linnæus*	85, 88
var. biaristatus	86
var. parviflorus	86
odoratus	91
Hordeum geniculatum	191
marinum	191
maritimum. *Withering*	187, 191
murinum. *Linnæus*	189

	PAGE
Hordeum nodosum	187
pratense. *Hudson*	187
rigidum	191
spurium	189
secalinum	187
sylvaticum. *Hudson*	185
Hydrochloa aquatica	103
Ischæmon vulgare	219
Koeleria cristata. *Persoon*	93
Knappia agrostidea. *Smith*	211
Lagurus ovatus. *Linnæus*	37
Leersia oryzoides	225
Lepturus filiformis	209
incurvatus. *Trinius*	209
var. filiformis	210
Lolium album	207
arvense	207
bromoides	149
italicum	205
multiflorum. *Lowe*	205
var. ramosum	205
var. submuticum	205
perenne. *Linnæus*	203
var. angustifolium	204
var. italicum	205
var. racemosum	204
var. tenue	204
rubrum	203
tenue	203
temulentum. *Linnæus*	203
var. longiaristatum	208
verum	207
Melica alpina	79
cærulea	79
decumbens	131
lobelii	83
montana	81
nutans. *Linnæus*	81, 83, 84
uniflora *Linnæus*	82, 83
Milium effusum. *Linnæus*	41
lendigerum	39
Mibora verna	211
Molinia cærulea. *Mænch*	79
var. breviramosa	80
var. depauperata	80

	PAGE		PAGE
Molinia depauperata	79	Poa cristata	93
Nardus stricta. *Linnæus*	5, 62	compressa. *Linnæus*	119
Ophiurus incurvatus	209	cæsia	125
Oplismenus crus-galli	97	distans. *Linnæus*	105
Panicum crus-galli. *Linnæus*	97, 101	var. minor	106
dactylon	217	var. obtusa	106, 110
glabrum	221	dulcis	65
humifusum	221	dubia	121
sanguinale	219, 221	decumbens	131
verticillatum	99	fluitans. *Scopoli*	107
viride	101	glauca	127
vulgare	97	glomerata	125
Pennisetum verticillatum	99	loliacea. *Hudson*	114, 117, 118
Phalaris alpina	31	laxa	126, 225
arenaria	21, 35	maritima. *Hudson*	106, 109, 110
arundinacea. *Linnæus*	21		
aspera	29	montana	127, 128
canariensis. *Linnæus*	19	nemoralis. *Linnæus*	127, 128
paniculata	29	var. angustifolia	128
phleoides	33	nutans	81
var.	21	pratensis. *Linnæus*	106, 115, 127, 128, 130
Phleum alpinum. *Linnæus*	27		
arenarium. *Linnæus*	21, 26, 31, 34, 35, 36	var. arida	116
		var. arenaria	116
asperum. *Jacquin*	29, 34	var. muralis	116
boehmeri. *Schrader*	31, 33	var. planiculmis	116
crinitum	45	var. retroflexa	116
commutatum	27	var. umbrosa	116
michelii. *Allioni*	26, 31, 34, 36	procumbens. *Curtis*	106, 110, 111
nodosum	25	polynoda	119, 128
paniculatum	29	parnelli	127
pratense. *Linnæus*	25, 26, 31, 34, 36	retroflexa	105
		rigida. *Linnæus*	113, 114, 118
var. longiciliatum	26	rupestris	111
var. longiaristatum	26	salina	105
viride	29	sylvatica	151
Phragmites communis. *Trinius*	223	subcompressa	119
Poa aquatica. *Linnæus*	103	scabra	121
annua. *Linnæus*	106, 129, 130	setacea	121
alpina. *Linnæus*	125, 126	subcœrulea	115
angustifolia	115, 127	trivialis. *Linnæus*	106, 121, 128
bulbosa. *Linnæus*	123		
balfourii	127	var. parviflora	122
borreri	225	trinervata	151

Polypogon littoralis. *Smith*	47
monspeliensis. *Desfontaines*	45, 48
Rottbœllia incurvata	209
Schedonorus elatior	145
sylvaticus	151
Sclerochloa distans	105
loliacea	117
maritima	109
procumbens	111
rigida	113
Serrafalcus arvensis	167
commutatus	165
mollis	169
secalinus	163
Sesleria cærulea. *Scopoli*	95
Setaria glauca. *Beauvois*	102
verticillata. *Beauvois*	99, 102
viridis. *Beauvois*	100, 101
Spartina alterniflora. *Loisel*	215
glabra	215
lævigata	215
stricta. *Smith*	213
Spartum anglicanum	23
austriacum	43
Stipa pennata. *Linnæus*	37, 43
membranacea	149
Sturmia minima	211
Syntherisma glabrum	221
vulgare	219
Trachynotia alternifolia	215
Trichodium caninum	55
Triodia decumbens. *Beauvois*	131
Trisetum flavescens	183
Triticum alpinum	197
biflorum	197
caninum. *Hudson*	196, 197
cristatum	226
junceum. *Linnæus*	193, 195
laxum	226
loliaceum	117
pinnatum	201
littorale	195
repens. *Smith*	195
var. aristatum	196
sylvaticum	199
unilaterale	117
Vulpia bromoides	153
uniglumis	149

LIST OF AUTHORITIES.

Abbot.
Aiton.
Allioni.
Arduino.
Arnott.
Babington.
Bauhin.
Beauvois.
Boehmer.
Borrer.
Brown.
Braun.
Bull.
Camerarius.
Cavanilles.
Cullum.
Curtis.
Dalechamps.
Davies.
Deakin.
De Candolle.
Deering.
Desfontaines.
Dickson.
Dillenius.
Dillwyn.
Don.
Dryander.
Dunal.
Dumort.
Ehrhart.
Fries.
Gaudin.
Gaudichaud.
Gerarde.
Gesner.
Godron.
Gouan.
Graves.
Greville.
Hall.
Haller.

Hoenke.
Hoffmann.
Hooker.
Hudson.
Hull.
Jacquin.
Jussieu.
Knapp.
Koch.
Koeler.
Kunth.
Leers.
Leysser.
Lightfoot.
Lindley.
Link.
Linnæus.
Loefling.
Lobel.
Loisel.
Lowe.
Macreight.
Mant.
Marschall.
Martyn.
Matthiolus.
Mitten.
Mœnch.
Morison.
Muhlemb.
Nees.
Nuttall.
Oeder.
Parlatore.
Parnell.
Persoon.
Petiver.
Plukenet.
Poiteau.
Pollich.
Pourret.

Purton.
Ralfs.
Ray.
Reichenbach.
Relhan.
Retzius.
Roemer.
Roth.
Rudbeck.
Salisbury.
Scheuchzer.
Schrader.
Schreber.
Schultes.
Scopoli.
Sibthorp.
Sesler.
Sidebotham.
Sinclair.
Smith.
Solander.
Sowerby.
Stillingfleet.
Sturm.
Thuill.
Towns.
Turpin.
Vahl.
Villars.
Wade.
Wahlenberg.
Watson.
Weber.
Weigel.
Wiggers.
Willdenow.
Wilson.
Winch.
Withering.
Woods.
Wulfen.

www.ingramcontent.com/pod-product-compliance
Lightning Source LLC
Chambersburg PA
CBHW032025220426
43664CB00006B/367